HANDBOOK OF EXPERIMENTAL METHODS FOR PROCESS IMPROVEMENT

HANDBOOK OF EXPERIMENTAL METHODS FOR PROCESS IMPROVEMENT

DAVID DRAIN

Intel Corporation, Chandler, Arizona

CHAPMAN & HALL

I(T)P® International Thomson Publishing

New York • Albany • Bonn • Boston • Cincinnati • Detroit • London • Madrid • Melbourne
Mexico City • Pacific Grove • Paris • San Francisco • Singapore • Tokyo • Toronto • Washington

Cover design: Trudi Gershenov
Cover photo: ©1997 PhotoDisc, Inc.

Printed in the United States of America

For more information, contact:

Chapman & Hall
115 Fifth Avenue
New York, NY 10003

Chapman & Hall
2-6 Boundary Row
London SE1 8HN
England

Thomas Nelson Australia
102 Dodds Street
South Melbourne, 3205
Victoria, Australia

Chapman & Hall GmbH
Postfach 100 263
D-69442 Weinheim
Germany

International Thomson Editores
Campos Eliseos 385, Piso 7
Col. Polanco
11560 Mexico D. F.
Mexico

International Thomson Publishing - Japan
Hirakawacho-cho Kyowa Building, 3F
1-2-1 Hirakawacho-cho
Chiyoda-ku, 102 Tokyo
Japan

International Thomson Publishing Asia
221 Henderson Road #05-10
Henderson Building
Singapore 0315

1 2 3 4 5 6 7 8 9 10 XXX 01 00 99 98 97
Library of Congress Cataloging-in-Publication Data

   Drain, David.
       Handbook of experimental methods process improvement / David
   Drain.
           p.   cm.
       Includes index.
       ISBN 0-412-12701-6 (alk. paper)
       1. Production engineering--Statistical methods.   2. Process
   control--Statistical methods.   3. Experimental design.   I.   Title.
   TS176.D73   1997
   658.5'072--dc20
                                                        96-34593
                                                          CIP

British Library Cataloguing in Pubication Data available

"Handbook of Experimental Methods for Process Improvement " is intended to present technically accurate and authoritative information from highlyl regarded sources. The publisher, editors, authors, advisors, and contributors have made every reasonable effort to ensure the accuracy of the information, but cannot assume responsibility for the accuracy of all information or for the consequences of its use.

To order this or any other Chapman & Hall book, please contact **International Thomson Publishing, 7625 Empire Drive, Florence, KY 41042.** Phone: (606) 525-6600 or 1-800-842-3636.
Fax: (606) 525-7778. e-mail: order@chaphall.com.

For a complete listing of Chapman & Hall's titles, send your request to **Chapman & Hall, Dept. BC, 115 Fifth Avenue, New York, NY 10003.**

To my parents Shirley and Donald Drain,
on the occasion of their fiftieth wedding anniversary

CONTENTS

PREFACE

This book was written to provide working engineers and engineering students with the statistical tools they must understand to improve manufacturing processes.

Statistical principles are illustrated with authentic manufacturing process examples from the semiconductor industry, a practice that enables experienced engineers to build upon existing knowledge to learn new skills, and which introduces students to a fascinating industry. The first book in this series, *Statistical Methods For Industrial Process Control,* gives background on the industry, and provides statistical understanding prerequisite to applying the methods in this text.

This text emphasizes the application of statistical tools, rather than statistical theory. Modern advances in statistical software have made tedious computations and formula memorization unnecessary, so engineers with knowledge of a few statistical tools can competently practice statistics within the context of their profession.

Software use is demonstrated throughout the book to promote better understanding through graphical and statistical analysis. A statistical software appendix gives example Statistical Analysis System (SAS)[1] programs sufficient to perform the analyses in the text. Some examples are shown with Minitab[2] as well.

Review problems at the end of each chapter give readers a chance to deepen their understanding. Answers to selected problems can be found at the end of the book.

[1]SAS is a trademark of SAS Institute Inc., SAS Campus Drive, Cary, NC 27513.
[2]Minitab is a trademark of Minitab, Inc., 3081 Enterprise Drive, State College, PA 16801.

A basic proficiency in algebra will be necessary to apply the apply the tools presented here. Calculus is required to understand the derivation of some techniques or underlying theory, but sections requiring calculus can be skipped without hampering effective statistical practice.

Chapter 1 is a nontechnical introduction to design of experiments (DOE)—it lays out some basic principles and serves to introduce following chapters. This chapter may be understood without reference to the remainder of the book.

Chapter 2 explains simple comparative experiments and introduces analysis of variance (ANOVA), which is used to analyze most types of industrial experiments.

Chapter 3 introduces the concept of blocking, and shows how it may be used to design more efficient and cost-effective experiments.

Chapter 4 reveals another time and money saver—the factorial experiment. This technique allows for the simultaneous investigation of several experimental factors.

Chapter 5 expands the subject of Chapter 4 to more general screening experiments like the fractional factorial and Plackett-Burman designs.

Chapter 6 is an introduction to optimization experiments; the method of steepest ascent and the simplex method are demonstrated.

ACKNOWLEDGMENTS

I would like to thank the many people who helped me write this book. Dave Martinich, Russ Sype, Russell Miller, and Shawn Clark gave extensive technical and editorial advice, which significantly improved the quality of the text.

A number of other people supplied examples, read portions of early manuscripts, or provided other support: Adrienne Hudson, Andy Gough, Ann Tiao, Anne Russell, Bill Connor, Blake Sacha, Brad Houston, Carl Memoli, Carlos Corral, Charla Frain, Chris Riordan, Chris Sluder, Chris Teutsch, Cindy Isabella, Curt Engelhard, Dale Brown, Dan Wiede, Daryl Lambert, Dwayne Pepper, Ed Bawolek, Eric St. Pierre, Erik Gillman, Fadi Geagea, George Stavrakis, Georgia Morgan, Gerard Vasilatos, Graydon Bell, Greg Headman, Harry Hollack, Janice Wittrock, Jim Moritz, Joan Hamre, John Ramberg, Julie Endress, Karl Einstein, Kelly Blum, Kevin Kurtz, Kurt Johnson, Lora Fruth, Lori Gates, Mark Johnson, Matt Gerdis, Matthew Ploor, Mike Bowen, Myron Weintraub, Neil Poulsen, Ralph Sabol, Ray Vasquez, Rita Dittburner, Rob Gordon, Rod Nielsen, Ron Gonzalez, Sharen Rembelski, Stan Mortimer, Steve Eastman, Susan Strick, Steve Thompson, Terri Rottman, Tim Lane, Tom Warner, Walt Flom, Warren Evans, and Wendell Ponder.

HANDBOOK OF EXPERIMENTAL METHODS FOR PROCESS IMPROVEMENT

CHAPTER 1

INTRODUCTION TO EXPERIMENT DESIGN

1.1 INTRODUCTION

Back in November I bought a new bicycle to drive to work, and I wanted that ride to be as comfortable and safe as possible. I know from my experience with previous bicycles that tire pressure has a lot to do with both of these important considerations: if the tires are overinflated the ride will be too bumpy, but if they are underinflated, my ability to corner accurately will be significantly impaired.

Determining the optimum tire pressure on the basis of scientific principles may have been possible, but it would have been difficult. I would have had to find equations relating tire pressure and tire material elasticity to this particular bicycle's response to bumps and turns; I would have had to quantify the responses I wanted to optimize (comfort and cornering), and I would have had to find some mathematical method to optimize these responses over the range of conditions in which I plan to ride.

Rather than engage in that research project, I chose to do an experiment: I inflated the tires to the pressure suggested in the bicycle manual, and then rode it for a few blocks around the neighborhood.

The ride was comfortable enough—the little pebbles and cracks I encountered did not cause excessive bumping, but my turning radius on that first trip was much too great to allow for safe and effective cornering. I decided that the tire pressure should be increased.

I pumped up the tires another 5 pounds and took the same route I had taken on the first ride. The bicycle did bounce and bump a little more, but it did not seem excessive. My ability to corner was much better, and certainly seemed adequate for the 8-mile drive I planned to take every day. I also noticed an added benefit: I was able to achieve higher speed with the same effort.

These two experiments were sufficient to arrive at a specification for tire pressure (40 pounds per square inch) under the conditions of the ride: riding at my present weight, with an empty basket, in Arizona at a temperature of about 70° on mostly level streets in good condition. If these conditions change significantly, or if I notice any degradation in the quality of my ride, I will consider changing tire pressure.

My bicycle experiments were similar to industrial experiments in several ways:

- I performed a sequence of experiments, rather than trying to get all of the information I needed out of a single big experiment.
- Even though I might have been able to find an optimum tire pressure purely through the application of scientific principles, it would have been very difficult, and I would not have had much confidence in the results.
- In the course of experimentation I discovered another important response (speed) which I had overlooked in the initial planning process. I will consider this response in any future experiments.
- The entire experimentation process was fast, and I have confidence in the results because I verified them under realistic operating conditions.
- The experimentation procedure is simple enough to repeat when the conditions of use change—when I lose 20 pounds and the temperature is 120°, I can just ride around the block a few more times to determine the optimum tire pressure.

An experiment is a deliberate manipulation of a process that intends to measure the effect of one or more experimental factors on some set of responses. Experimental factors may be (among others) machine settings, starting material, operating procedures, or environmental conditions. Responses might be polysilicon resistivity, yield, particle counts, or any other measurable process result of economic or scientific interest. A factor has an effect on a response if different levels (settings) of that factor produce a change in the response (Figure 1.1).

Experiments provide knowledge necessary to improve or repair processes. Experiments can find recipe changes that will move a process mean to target; they can isolate sources of defects and test methods for their elimination; they can assess measurement capability and optimize measurement procedures; and they can help apply new machines, methods, or materials to economic advantage in an existing process.

Response

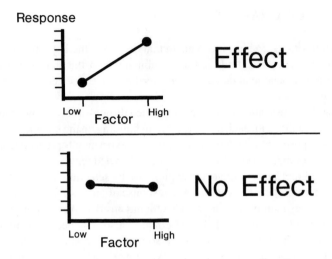

Fig. 1.1   A factor has an effect on a response if different levels (settings) of that factor produce a change in the response. In an experiment represented by the top half of this figure, the factor (*x*-axis) was set to two levels, low and high. The response (*y*-axis) took a different value at the low setting than at the high setting, so the factor has an effect on the response. In another experiment (bottom half of the figure), the factor has no effect on the response because changes in the factor do not provoke corresponding changes in the response.

Experiments are the most expedient medium through which this knowledge may be obtained: passive observation of a stable process yields information only on the process as it is, not as it could be; and in spite of modern scientific developments and the ready availability of computing power for simulation, science and engineering knowledge are inadequate to faithfully represent complex processes like thin-film deposition or plasma etching.

Experiment design is a discipline based on sound statistical principles and built on years of previous engineering experience; its application produces experiments that are effective and expedient in meeting their stated goals. Experiment design helps determine the factors and responses in an experiment, factor settings, resource requirements, and the manner in which the experiment is performed.

The main benefits of designed experimentation are that critical information is obtained faster, more economically, and more reliably than it would be with haphazard or naive experimentation. Designed experiments do have risks—of erroneously concluding a factor has an effect on a response, for example—but those risks are known and quantified before the experiment is conducted, and design principles can be used to keep risks within reasonable bounds. Undesigned experiments have much more serious risks—that of mistaking the influence of one factor for another, for example—and the probability of those risks cannot be assessed.

1.2   PROCESS CHARACTERIZATION

Process characterization is a mechanism that discovers the important factors acting on process outcomes, finds setting combinations of the factors that are most likely to produce beneficial outcomes, and verifies that the selected recipe satisfies factory needs.

Process characterization is best accomplished through a set of experiments conducted sequentially, with each new experiment building on the knowledge obtained in previous experiments. As shown in Figure 1.2, experiments proceed from a state of relative ignorance to one of more complete knowledge.

Screening experiments initiate the exploration by selecting active factors (those that produce a significant effect on some response of interest) from many possible candidates—they mainly determine if a factor has an effect on a response, although some estimate of the magnitude of the effect is also obtained. Screening is performed early in the experiment sequence so that very little time is spent investigating unimportant factors. Screening experiments are also effective trouble-shooting tools.

Once active factors have been identified, *optimization experiments* are used to find the most advantageous factor settings, even if tradeoffs between several responses must be taken into account. Screening experiments often investigate 10 or more factors; optimization experiments rarely include more than five.

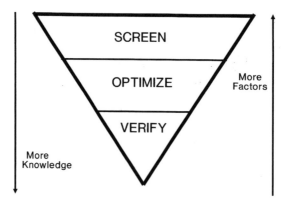

Fig. 1.2   Process characterization proceeds from a state of relative ignorance (base of the triangle) when many variables are considered as candidates for having a significant effect on responses of interest. Screening experiments find the few truly important factors that are used in optimization experiments to select process recipes. Verification experiments (tip of the triangle) involve only one or two factors, and are meant to provide evidence that the chosen process recipes will meet factory needs in the high volume manufacturing environment.

Verification experiments are used to prove that newly modified processes do indeed perform as expected in a manufacturing environment. They are often single-factor experiments—a simple comparison between one recipe and another—but they obtain precise data on differences in process performance.

1.3   THE PROCESS OF EXPERIMENT DESIGN

Before an experiment can be designed, several important prerequisites must be satisfied. First, measurement capability for responses and factors must be sufficient for the needs of the experiment. Poor response measurement capability requires the experimenter to expend relatively large effort and expense; poor factor measurement capability can lead to deceptive results.

Second, processes that produce raw materials for the experiment should be stable, as should processes influencing experimental material after the experiment and before response measurements. Extraneous variation from unstable processes makes it difficult to measure the effects of experimental factors; more wafers are required to find differences, so experiment cost increases.

Third, resources sufficient to obtain useful results must be committed to the experiment: experimental material, equipment time, metrology resources, and the people to process material, make measurements, and analyze results are all necessary to the success of an experiment.

Lastly, the right people must be involved in the design process from the beginning. Engineers with some historical knowledge of similar situations can envision responses that should be measured, restrictions that must be imposed on factors, and possible side effects of anticipated process changes. Those who will execute the experiment in the factory will notice logistics considerations that affect the manner in which the actual experiment will be executed. Statisticians can point out shortcuts and prevent pitfalls before the experiment, but even the most clever statistical expert cannot repair data from a poorly designed experiment.

The experiment design process includes the following steps:

1. Define the purpose and scope of the experiment.
2. Examine scientific literature and documentation from previous experiments.
3. Choose experiment responses.
4. Choose experiment factors and levels.
5. Account for other experimental variables.
6. Choose the experiment structure.
7. Determine experiment risks and resource requirements.
8. Outline experiment execution.
9. Plan data analysis and decision making.

These steps are discussed in detail below.

Active factor: Active factors produce an effect on some response of interest.

Blocking factor: A blocking factor is a source of response variance that adds or subtracts some offset to every experimental unit in the block. This generally causes experimental units within the block to be more similar to one another than they are to experimental units in other blocks.

Categorical factor: Experimental factors that can take only a finite number of values on a nominal scale are categorical.

Confounding: The unfortunate experimental situation wherein two factors simultaneously vary together throughout an experiment. Their effects cannot be untangled, so the factors are said to be confounded.

Continuous factor: Experimental factors that can take any of an infinite number of values within their range are continuous.

Controlled variable: A variable is controlled if it is forced to take the same value for the course of the entire experiment.

Economic response: Economic responses are measured because of their importance to company profit. Even though they may not actually be affected by experiment factors, they must be included in experiments because many process changes may have unexpected and unpleasant side effects; these are better anticipated before a process change is made rather than corrected after a disaster.

Effect: A factor has an effect on a response if different levels of that factor produce a change in the response.

Experiment: An experiment is a deliberate manipulation of a process which intends to measure the effect of one or more experimental factors on some set of responses.

Factor: Experimental factors are deliberately manipulated to determine their effect on responses of interest.

Factor setting: A particular value at which a factor is fixed for part of an experiment.

Figure of merit: A figure of merit is a complex assessment of overall process performance—often a mathematical combination of more straightforward responses.

Indicative response: Indicative responses are especially sensitive to the influence of experimental factors. They are most likely to give the experimenter valuable information about what affects the process, and what settings might best achieve process parameter targets.

Level: A particular value at which a factor is fixed for part of an experiment. Even though this term seems to apply only to continuous variables, it is often applied to categorical variables.

Noise factor: Noise factors can be controlled in experiments, but cannot be controlled in normal manufacturing. Including these factors in experiments and studying their effects on responses can lead to more robust processes.

Optimization experiment: Optimization experiments are used to find the most advantageous combination of factor settings, even if tradeoffs between several responses must be taken into account.

Process characterization: A learning process wherein the important factors acting on process outcomes are discovered, setting combinations of the factors which are most likely to produce beneficial outcomes are found, and recipes that satisfy factory needs are selected and verified.

Randomization: The practice of randomly ordering the combinations of factor settings to be run in the experiment. Note that the combinations of factor settings are carefully determined; only the order in which they are run is random.

Replication: The practice of repeating an experiment to gain further confidence in the results and provide a basis for statistical tests.

Response: Responses are measurable experiment results of economic or scientific interest.

Risks: Experiment risks are of two types. A Type I error occurs when the experimenter falsely concludes that a factor has an effect, when in fact it does not. The probability of Type I error is denoted by the Greek letter α, and the risk is termed *alpha risk.*

A Type II error occurs when the experimenter fails to detect a factor effect when it is truly present. The probability of Type II error is denoted by the Greek letter β, and the risk is termed *beta risk.*

Robust process design: Robust process design is a modern innovation in experiment design which can be used to create more robust processes—ones that consistently produce excellent products in spite of aberrations in raw materials, process variation from previous steps, and habitual differences in processing due to people and machines.

Scope of experimentation: The scope of experimentation determines the extent to which experiment results can be generalized. Exercising a diversity of machines, settings, environmental conditions, and materials in an experiment enhances the ability to generalize results to broader circumstances.

Screening: Screening experiments simultaneously evaluate the effects of many different factors. Their primary purpose is to discern active factors from inactive factors, although they may also provide some estimates of factor effects.

Side-effect: An unintended influence on some response due to a change in an experiment factor.

Significant effect: An effect is statistically significant if a hypothesis test rejects the null hypothesis; an effect is practically significant if the size of the effect is great enough to be of economic importance. Some statistically significant effects are not practically significant.

Uncontrolled factor: A factor, known or unknown, that is allowed to vary freely throughout an experiment.

Verification experiment: Verification experiments are used to prove that newly modified processes do indeed perform as expected in a manufacturing environment. They are often single-factor experiments—a simple comparison between one recipe and another.

1.   Define the purpose and scope of the experiment.

Good experiment designs obtain the exactly the information required in the most expedient and economical manner, so experiment goals must be clearly understood. A statement of purpose and scope should answer the following questions:

- What situation will be improved with information from this experiment?
- What decisions will be based on the results of this experiment?
- How may the results be generalized?

Most experiments can be assigned to one of the purpose categories in Figure 1.2: screening, optimization, or verification. One of the gravest errors an experimenter can make is to confuse two different purposes into one experimental goal—this invariably wastes experimental material and impedes learning. No single experiment is sufficient to completely characterize a process—each individual experiment should be considered within the wider context of the sequence of process characterization experiments.

The scope of experimentation should also be explicitly defined: is only one piece of equipment of interest? or all pieces of similar equipment? or any equipment of similar capability? Exercising a diversity of machines, settings, environmental conditions, and materials in an experiment enhances the ability to generalize results to broader circumstances, but this adds complexity and cost to the experiment. Unless such generalization is desirable, the added cost is superfluous.

2.   Examine scientific literature and documentation
from previous experiments.

The design process should include comprehensive background research into the chemical and physical aspects of the process under study, and a review of prior experimentation. This inquiry may suggest experimental factors and reasonable ranges over which they should be varied; it may eliminate some suspected factors or interactions; in some cases, it may even prove that further experimentation would be unnecessary or fruitless.

Be critical of results from prior experiments, especially if they are not verified by independent experimentation, or if their conclusions conflict with scientific principles. Previous research may have been flawed by poor experimental practice or erroneous interpretation.

3. Choose experiment responses.

Most experiments should have at least two different types of responses. Some responses are chosen because they are expected to be especially sensitive to the experimental factors. These *indicative responses* are the most likely to give the experimenter valuable information about what affects the process, and what settings might best achieve process parameters targets. For the gate oxide process, gate oxide thickness is a good indicative response—nearly any change to a process recipe is likely to affect it.

Some responses are chosen for economic reasons—die yield for example. Any experimental factor level which had a large adverse impact on yield would obviously be an unacceptable choice for a process setting. These *economic responses* are a necessary part of experimentation because many process changes may have unexpected and unpleasant side effects; these side effects are better anticipated before a process change is made rather than corrected after a disaster. Other common economic responses may be safety or environmental considerations, particles, device reliability, equipment or materials cost, process throughput time or logistics, machine wear, and measurement time.

A response may be a complex assessment of overall process "good performance," or a *figure of merit.* For the gate oxde process, the following quantity is one possible figure of merit:

$$M = \frac{1}{(\text{Thickness} - \text{Target})^2} - \text{Particles}$$

Decreasing thickness differences from target increases M, as does particle reduction.

The measurement precision of responses must be known before any conclusions can be reached about the effects of experimental factors upon them. While this may seem perfectly obvious, one of the most common errors in experimentation is to attempt to determine the influence of factors on unmeasurable responses.

4. Choose experiment factors and levels.

Experimental factors are those process variables that are deliberately manipulated in an experiment to see how they affect a response of interest. Some factors are categorical (or qualitative), like the brand of quartz used in the gate oxide process. Suppose there are only two quartz vendors with sufficient supply, then the only possible levels or settings for quartz vendor are brand W and brand V. If another quartz vendor were qualified, then the brand factor could have a third level.

Experimental factors can also be continuous or quantitative; oxidation temperature would be a continuous factor in experiments on the gate oxide process. Settings for continuous factors must be chosen—they are not predetermined as they are for categorical variables. They should be chosen so that useful information is obtained as each setting: setting oxidation temperature to 30°C would be useless because a negligible amount of silicon dioxide would grow at that temperature; setting the temperature to 1800°C would be useless because the wafers would melt. Settings must also be chosen so that the effect of the factor will be evident if it is present: choosing only temperatures of 600 and 602°C would probably not achieve this end because the responses at these settings would be indistinguishable.

Experiments can have more than one experimental factor. The number of experimental factors depends mainly on the stage of characterization: screening experiments have many factors, and verification experiments have few.

5.   Account for other experimental variables.

Experimental factors are not the only influences on responses: blocking factors, noise factors, deliberately controlled (constant) variables, and uncontrolled variables are also potential influences. Each of these must influences be identified and accounted for if the experiment is to be effective.

A *blocking factor* is a source of response variance that adds or subtracts some offset to every experimental unit in the block. This generally causes experimental units within the block to be more similar to one another than they are to experimental units in other blocks. For the gate oxide process, a batch of wafers processed through the same diffusion tube at the same time form a natural block: all the small differences unique to that particular furnace and run influence the wafers as a group. Blocking factors occur naturally, they cannot be avoided, and they add variance to experiment responses.

A variable is controlled if it is forced to take the same value for the course of the entire experiment. For the gate oxide process, reaction pressure is one possible choice for a controlled variable: if the standard recipe requires a pressure of 200 mtorr and there is no reason to change this, then fixing pressure at 200 mtorr will prevent pressure from causing changes in gate oxide responses (thickness, for example) that could obscure the effects of more interesting factors like temperature or reaction time.

Variables that can be controlled only during experiments, or that are difficult or expensive to control during normal production, are called *noise factors*. Load size (number of wafers in the tube) is a noise factor for the gate oxide process—it is very simple to control during an experiment, but impractical to precisely control during normal production. Noise factors are included in experiments to assess process robustness—the immunity to differences in environment, raw materials, and other common manufacturing variations.

Uncontrolled variables are allowed to vary freely during an experiment, although they may be measured and recorded. Variables may be left uncontrolled because they almost surely have no influence on any response of interest. For example, lot box color probably has no effect on gate oxide thickness, so it would not be controlled during experiments with this response. Other variables are involuntarily left uncontrolled because they are difficult or impossible to control: atmospheric pressure, for example.

Some variables are uncontrolled because they are unmeasurable, or accidentally ignored. If such a variable happens to change in unison with an experiment factor, it can make that factor appear to be more or less influential than it truly is. When this happens, the uncontrolled variable is said to be *confounded* with the factor. If, for example, reaction pressure were allowed to vary in a gate oxide experiment, and three reaction temperatures (1100, 1120, and 1140° C) were tried in order, then any resulting change in gate oxide thickness could be due to temperature, or to pressure, and unless pressure were measured throughout the experiment, it would be impossible to tell which factor caused the differences.

Too much control in experiments is as deleterious as too little control. If a process is controlled much more closely during an experiment than it would be in normal operation, results might not generalize to the process—responses could be acceptable during the experiment, but disastrous on the first day of normal manufacturing. Some ways to achieve an acceptable level of control are to:

- Make sure that processes making raw materials for the experiment are stable. Instability in incoming materials translates to extraneous variance in responses, thus hiding the effects of experimental factors. Because well-managed processes are usually stable, evaluating factor effects in the presence of excess incoming materials variation is of little interest.
- Make sure that processes following the experiment but preceding response measurements are also stable. Some responses are measured long after experiments (electrical parameters, for example), so extra variation induced after the experiment will hide the effects of factors.
- Replicating an experiment over blocking factors like diffusion run, shift, or time, will ensure some degree of generalizability to the broader population of manufactured product.
- Controls that are restrictive compared to the typical process should be carefully documented so that corresponding restrictions on generalizability will be recognized.

6. Choose the experiment structure.

Certain standard types of experiments have been found to be especially effective in satisfying particular experiment goals. Table 1.1 shows the correspondence between experiment goals and experiment structures that have been found to best satisfy those goals. Using standard experiment structures simplifies the design process—a few basic patterns can be combined and embellished to suit most situations, and each of the suggested experiment types is sufficiently adaptable to accommodate blocking, control, and other categories of variables.

7. Determine experiment risks and resource requirements.

Every experiment has risks—an incorrect conclusion could be obtained and a damaging decision made based on that conclusion. One of the most powerful as-

TABLE 1.1   Summary of Experiment Applications

Activity	Experiment Type	Chapter				
		2	3	4	5	6
Screen • Troubleshoot • Investigate • Find noise	Fractional factorial Plackett-Burman Factorial			✓	✓	
Optimize • Find target • Improve yield • Reduce cost • Make robust	Central Composite Simplex Factorial			✓		✓
Verify • Match • Find variance sources	Comparative Blocked experiments*	✓	✓			

*Experiment types vary according to their purposes: early process characterization requires screening experiments; optimization experiments follow; and verification experiments are performed on relatively mature processes.

pects of modern experiment design is that risks are known and limited before the experiment is performed. Risks often stated in terms of the two types of error which can be made when testing hypotheses:

- *Type I error.* Falsely concluding that a factor has an effect, when in fact it does not. The probability of Type I error is denoted by the Greek letter α, and the risk is often termed alpha risk.
- *Type II error.* Failing to detect a factor effect when it is truly present. The probability of Type II error is denoted by the Greek letter β, and the risk is often termed beta risk. Beta risk is different for each size of effect (big effects are easy to detect), so it must always be associated with a specific difference, or alternative.

■ *Example 1.1: Experiment Goals*

Suppose a change to the gate oxide recipe were being considered. The new recipe is faster, but requires a higher temperature for most of the diffusion run. If the new recipe were found to produce gate oxide at the same average thickness as the present recipe, the process would be changed immediately. Experiment goals might be stated as follows:

This experiment must detect a gate oxide thickness shift of 12 Å 90% of the time, but falsely indicate the existence of a recipe effect only 5% of the time.

This means that alpha risk is 0.05, and beta risk is 0.10 with regard to a specific alternative of 12 Å.

Choosing acceptable experiment risks is easiest if the consequences of each type of error are considered. For the gate oxide recipe change experiment, a Type I error would preclude a beneficial recipe change and waste processing time for all future gate oxide runs. A Type II error would allow the recipe change to be implemented even though it would often produce gate oxide thicknesses at least 12 Å away from target.

Once acceptable experiment risks are explicitly stated, the resources necessary for the experiment can be determined. Cost and risk conflict: experiments with acceptable risks may seem too costly, and experiments with reasonable costs often seem too risky. Some rational compromise between the two is usually necessary. Try relaxing the experiment risk constraints—maybe finding a 15 Å (rather than 12 Å) difference in gate thickness 90% of the time is sufficient for a process with a 525 Å thickness target. One of the most common errors of experiment design is to try to find differences that are not practically significant.

Another alternative is to reconsider cost constraints: perhaps more time and money should be spent because of the potential consequences of Type I or Type II errors. Financial professionals can be called upon to quantify costs associated with false experiment conclusions—these costs are often greater than they might first seem.

If no compromise seems possible and an experiment with maximum acceptable risk is still too costly, no experimentation should be undertaken. Incomplete information obtained from miserly experimentation will probably mislead the experimenter by giving a false sense of security and lead to long-term losses greater than the cost of a responsible experiment.

8.   Outline experiment execution.
One trivial execution error can ruin an otherwise superbly planned experiment and waste all the time and materials that were dedicated to the experiment, or even worse, yield deceptive results leading to devastating consequences. Every detail of experiment performance and logistics must be planned and checked with the people who will actually do the experiment.

Obtain materials and secure access to equipment before the experiment. Find out who will perform any critical or unusual parts of the experiment (extra measurements, recipe changes on machines) and obtain the commitment of workers and their supervisors. The surest way to enlist people in the experiment is to assure them they will be informed of the results—curiosity is a more powerful motivator than overtime pay.

Use instruction sheets, checklists, and data collection forms for any aspect of the experiment that differs significantly from normal production. Write clear instructions that leave nothing to chance: if diffusion tube 4B must be used for an experiment, denote this fact explicitly. Experimentation often requires people to do things they would not do during normal production, so comprehensive directions are absolutely necessary. Have an operator read through the instructions before the experiment, and make sure they understand them.

Make the experiment as simple as possible: include only those factors and treatments which are necessary, and do not request unnecessary measurements. Complex experiments are more prone to execution error than simple ones, and they depend more on perfect execution to achieve their goals.

Experimental material is always at greater risk than normal production material just because it is unusual, so lessen the risk of errors by making the experiment look as much like normal production as possible: use the standard lot size, the same boats, and the same measurement equipment. Results from experiments processed in this manner are also more easily generalized to standard production.

Design redundancy and robustness into the experiment itself so that is able to tolerate a reasonable number of lost wafers and minor mistakes. Adding a few wafers to the design in high-risk parts of the experiment—where wafer breakage is likely, for example—helps ensure that at least some information is obtained from that part of the experiment. Never commit an entire experiment to a single diffusion run—that will be the one time a cat jumps into an electrical transformer and causes a power failure for the entire factory.

9. Plan data analysis and decision making.

Information obtained through analysis of data from experiments is used to make decisions, so these analysis methods and decision rules are just as important as the experiment itself. For example, if the goal of an experiment is to validate a process change, success criteria for the new process should be clearly defined—this prevents arguments about the prudence of the change and speeds the decision-making process.

Visualize data analysis before the experiment to make sure the planned analysis is rational and sufficient for decision making needs. Try analyzing simulated data before the experiment to check methods and anticipate the appearance of the final report.

The paradigm for experiment design presented here will be exercised thoroughly in following chapters. Those desiring more information on this design process should consult Coleman and Montgomery (1993).

1.4   THE INFLUENCE OF TAGUCHI

Robust process design is a modern innovation in experiment design that can be used to create more robust processes—ones that consistently produce excellent products in spite of aberrations in raw materials, process variation from previous steps, and habitual differences in processing due to people and machines. Recent interest in the field was sparked by Genichi Taguchi (Taguchi, 1987) who proposed a set of experiment designs and analysis methods for robust process design.

There has been some controversy concerning Taguchi's methods—see Pignatiello and Ramberg (1991) for details. In spite of the debate, at least three of Taguchi's beliefs are certainly worth adopting:

1. Keeping a process parameter within specification limits is not good enough—any deviation from target is harmful, and bigger deviations are more harmful than smaller ones.

A finished part with all important characteristics within specification limits but far from target is not as desirable to the customer as one with those characteristics exactly on target. A loss function is a gauge of loss as a function of the distance of a process parameter from target; loss could be measured in terms of cost of production, die yield, cost of customer returns, or throughput time, for example. Loss functions can themselves be used as experiment responses; by doing so, experiments should lead to processes that have the lowest average loss.

2. The variance of a response is at least as important as the average response when choosing process parameter settings for recipes.

If two processes are on target, but the first has higher variance about target than the second, the second process is superior to the first—it will produce more material with parameter values near target. This point is more profound than it appears at first. For most processes, it is much easier to change the process average than it is to reduce the variance. Thus, a stable process with small variance will tend to produce more material near target than one with high variance, and it will also be more easily repaired to correct occasional instability.

Experiments can exploit this concept by using the standard deviation of a response as a separate response, and then choosing recipes that minimize it.

3. There is a class of factors that can be controlled in experiments, but not under the conditions of high-volume manufacturing. Including these factors in experiments and studying their effects on responses can lead to more robust processes.

There are many aspects of production that vary in a relatively uncontrolled manner: the time since a piece of equipment had its last preventative maintenance, or the precise the number and positioning of wafers in a diffusion tube, for example. Any of these factors could affect important responses, but they are often ignored simply because they are difficult or uneconomical to control during normal production.

Taguchi referred to these as *noise factors,* and suggested that experiments should be designed explicitly to investigate their effects at potential process recipes. Those recipes that are robust to noise factor variation are superior recipes for normal production use because they will generate less variation from target.

Noise factors can also be treated like any other experimental factor—they can be deliberately varied, and their effects estimated. Noise factors causing large effects may be such significant sources of process instability that they must be eliminated, or their effects ameliorated in some way. For example, if the time since preventative maintenance were found to be the most important factor in determining gate oxide thickness, maintenance frequency could be increased to lessen the effect.

1.5   PREVIEW OF FOLLOWING CHAPTERS

The five chapters following this one describe and demonstrate experiment types appropriate for each step of process characterization. Experiments are presented in order from simplest to most complex so that the principles of good experimental design can be most easily learned; the order of experiment application in actual practice will be roughly backward to the presentation order.

Chapter 2 presents simple comparative experiments and introduces analysis of variance (ANOVA), the statistical tool that is used throughout experiment design to analyze results and make decisions regarding the importance of the effect of a factor on a response.

Chapter 3 introduces the concept of blocking—a design principle that ensures more robust and economical experiments. Some form of blocking is utilized in almost every industrial experiment, including screening and optimization experiments.

Chapter 4 introduces the multifactor designs known as factorial experiments. Factor interactions—effects that only appear when two factors are varied simultaneously—can only be understood with factorial designs, and in the semiconductor industry such interactions are common. Factorial designs also present impressive economic benefits because they allow for the investigation of several factors in a single experiment. Factorial experiments further quantify results from screening, and often obtain effect estimates sufficient for optimization.

Chapter 5 describes screening experiments that can sort through a large number of factors, and then select for further study those which have some effect on responses of interest. Fractional factorial and Plackett-Burman designs are presented.

Chapter 6 presents some optimization methods—experiments, or sequences of experiments, that discover ideal factor settings. The central composite design—an outgrowth of factorial experiments—is the main topic of this chapter. Simplex designs are also demonstrated; these general optimization experiments are effective with multiple predictors, even when no assumptions may be made about the functional relationship between the predictors and the response.

COMPARATIVE EXPERIMENTS

2.1 INTRODUCTION

The most frequently performed type of industrial experiment is one that compares the means of two or more populations. Comparative experiments are used to test pieces of equipment to see if they produce equivalent results, and to quantify the effects of different recipes or machine settings. They are also used to qualify process changes, that is, to ensure that a new process is equivalent to or better than the standard process.

This chapter introduces principles of good experimentation and methods of analysis that are used for most types of industrial experiments. Section 2.2 gives a precise description of comparative experiments and the statistical models used to interpret them. Section 2.3 explains how to design comparative experiments to achieve experiment goals as cheaply as possible, and Section 2.4 explains how to successfully execute comparative experiments. Section 2.5 is devoted to analysis methods, and also gives guidance on explaining analysis results to others.

Some experiments for which the usual set of distributional assumptions do not apply are presented in Section 2.6.

2.2 COMPARATIVE EXPERIMENTS

The purpose of a simple comparative experiment is to see if the means of a set of populations are essentially the same. This is not the same as comparing each mean individually against some standard; but rather the means of a set of populations are simultaneously compared to one another.

Consider a factory that has four polysilicon doping tubes. Poly resistivity is a critically important output of the doping process, and the mean resistivity from one tube concisely describes the quality of polysilicon produced by it. A simple comparative experiment can determine if the means of these four populations of poly resistivity are the same.

The consequences of an incorrect decision based on this type of experiment are serious: if the populations are different and the difference goes undetected, then an opportunity to reduce process variance is lost. If the populations are the same, but for some reason a difference is falsely perceived, engineering time will be wasted trying to fix tubes in good repair, and process variance is likely to increase because of this meddling.

The method used to choose between real and apparent differences is analysis of variance (ANOVA). ANOVA compares an estimate of the differences caused by an experimental factor (tube, in the example above) to an estimate of within-group (within-tube) variance. If the ratio of the two is sufficiently large, then the apparent difference is regarded as real.

The statistical test provided by ANOVA is actually necessary to make a decision. Suppose, for example, that an experiment was run in which 11 runs were made on each of four poly doping tubes, and that the mean resistivity obtained from each tube were as shown below:

Tube	13A	13B	13C	13D
Rho	26.0	29.1	24.7	31.0

Since these means are not all exactly the same, some people would say that the tubes must be different in some way.

Now, look at side-by-side boxplots of the 11 measurements from each tube (Figure 2.1). Even though the means are not the same, the inherent variability in poly resistivity is so great that the apparent difference in means is insignificant. Another experiment just like this one could produce means that seemed the same; or the means could seem different in some other way, for example, 13A greater than 13B, rather than less.

An entirely different situation is represented in Figure 2.2; here, the difference in means appears to be truly important compared to the intrinsic variability of poly

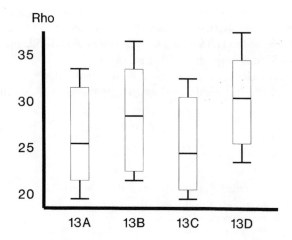

Fig. 2.1   Side-by-side boxplots of 11 measurements from each of four polysilicon doping tubes are shown. The inherent variability in poly resistivity is so great that the apparent difference in means is insignificant.

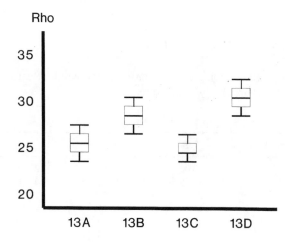

Fig. 2.2   Side-by-side boxplots of 11 measurements from each of four polysilicon doping tubes are shown. The inherent variability in poly resistivity is small compared to the difference in means.

resistivity in each tube. Another experiment exactly like this one might yield slightly different means, but the relationship between the means would remain essentially the same, and the differences between them would still appear to be significant compared to the inherent poly resistivity.

ANOVA is the statistical tool that can tell important differences from insignificant differences. The statistical model underlying this type of experiment is:

$$y_{ij} = \mu + \tau_i + \varepsilon_{ij}$$

where

y_{ij} is the response for the jth replicate at the ith level of the factor. y must be continuous, not categorical.

μ is the overall population mean.

τ_i is the effect of the ith level of the factor. This is the amount of resistivity added or subtracted whenever the factor is at level i. Because these τ_i are offsets from an overall mean, their sum must be zero.

ε_{ij} is a random error for the jth replicate at the ith level of the factor. These errors are assumed to be independent and normally distributed with mean zero and constant variance σ^2.

$i = 1 \ldots p$, where p is the number of levels of the factor.

$j = 1 \ldots n$, where n is the number of replicates at each level of each factor.

■ *Example 2.1: Statistical Model for a Poly doping Experiment*

Data for a poly silicon doping experiment is shown below. Resistivity from 11 runs in each of the four tubes was measured.

13A	13B	13C	13D
23.50	30.00	15.50	22.25
17.75	19.75	24.75	27.50
21.00	20.25	26.75	19.50
24.25	21.25	16.75	24.50
18.75	18.75	18.00	23.50
27.50	23.75	22.75	18.00
26.00	21.50	20.25	24.00
23.50	26.25	26.00	25.25
22.25	22.75	15.50	26.50
22.75	22.50	26.00	19.75
26.50	19.50	16.25	20.50

The ANOVA model is:

$$y_{ij} = \mu + \tau_i + \varepsilon_{ij}$$

where

y_{ij} is the resistivity for the for the jth run in the ith tube.

μ is the overall resistivity mean, that is, the average for the entire process.

τ_i is the effect of the ith tube. This is the resistivity added or subtracted by the each tube.

ε_{ij} is a random error for the jth run in the ith tube. These errors are assumed to be independent and normally distributed with mean zero and constant variance σ.

$i = 1 \ldots 4$, because there are four tubes in the experiment. The tubes are named 13A, 13B, 13C, and 13D.

$j = 1 \ldots 11$, because 11 runs (replicates) were done in each tube.

This model is similar to linear regression models, with one important difference: linear regression models have one or more continuous factors; here factors can be continuous or categorical. The poly tube factor is categorical—tubes are distinctly different things, not measurements along some scale. Continuous factors are generally better understood through the use of linear regression models than with the ANOVA model above, so most of the following discussion will focus on experiments with categorical factors.

The model shown here is a fixed-effect model, so called because factor levels can be exactly duplicated over replications: tube 13A is always exactly tube 13A. Random effects are those for which factor levels are randomly selected from a large population, and the factor is considered a variance component.

The model statement in a variance components model looks like a fixed-effect model, but each τ_i is a random sample from a normally distributed population with mean zero and fixed variance, σ_{factor}. For one-factor experiments, this distinction makes no difference in the ANOVA calculations, so for the remainder of this chapter a fixed-effect model will be used.

2.3   DESIGN

Comparative experiments are relatively easy to design, but there is still plenty of opportunity to make mistakes that can reduce the effectiveness of the experiments, unnecessarily increase their cost, or even lead to false conclusions. To help avoid these problems, the design paradigm provided in Chapter 1 will be used here to describe sound design practices for comparative experiments.

Step 1: Define the purpose and scope of the experiment.

Simple comparative experiments all have essentially the same purpose: to determine if a factor has an effect on a response. In the poly doping context, a comparative experiment could be used to determine if poly resistivity is significantly affected by the diffusion tube that is used during doping.

Extrapolation is just as dangerous in comparative experiments as in linear regression, so the conclusions of an experiment on a (fixed) categorical factor should only be applied to those factor levels included in the experiment. Results from experiments involving continuous factors (temperature, for example) can be usually be applied to any value within the range of the original factor levels—this is interpolation, not extrapolation.

For the poly doping experiment, the scope is the set of tubes included in the experiment; no inferences can be made about other tubes, or about conditions of operation different than those prevailing during the experiment. The conclusions obtained in this experiment could not be applied after recipe changes, raw material changes, or any other alteration of the usual process. If any of those other conditions were of interest, experiments would have to be designed to investigate them.

Experiments should only be conducted when the process is stable, so conclusions would apply only to the stable process. Special causes that are active during process excursions fundamentally alter the process—it is in essence a different process during excursions than when stable.

Step 2: Examine scientific literature and documentation from previous experiments.

The foremost reason for reading old experiment reports, process logs, or other history is to avoid repeating mistakes. The errors of previous experimenters may also be revealed in old reports, and thus avoided in future experiments. Perhaps an unanticipated variable must be carefully controlled, or a measurement system has to be recalibrated every day during the experiment.

Another good reason for this research is to obtain an estimate of within-group variance. This estimate will be used to determine the experiment sample size, so the more accurate it is, the more cost-effective the experiment will be.

Step 3: Choose experiment responses.

In most comparative experiments, there is a clear choice for at least one indicative response—the one that originally motivated the experiment. For the poly doping experiment, poly resistivity would surely be measured.

In any experiment in the semiconductor industry, yield and particle counts would be examined as an economic responses, and there may be others like throughput time.

Make sure measurement equipment has sufficient measurement capability to detect important differences in any of the chosen responses.

Step 4: Choose experiment factors and levels.

By definition, simple comparative experiments have one factor. Factor levels should be chosen so that useful information can be obtained at all levels. In the poly doping experiment, it would not make sense to include a tube that never runs poly dope, and is not intended to ever run poly dope. Any measurements obtained from that tube would have no relevance to the usual process.

Levels for continuous factors can be more difficult to select because there are an infinite number of choices, but the same principle applies: factor levels should be within a range likely to produce useful information. For example, poly doping is known to occur only at temperatures over 550°C, so including a level of 500°C would only waste material. Continuous factors should usually have at least three equally spaced levels to facilitate linearity checking.

Step 5: Account for other experimental variables.

Simple comparative experiments investigate the effect of a single factor, so other sources of variance should be avoided. Tube time and temperature are known to have an important effect on poly resistivity, so these should surely be controlled in an experiment where tube is the only factor.

Some variables that are not normally controlled may need to be controlled in an experiment. Choosing a particular measurement system, a particular operator, or a single batch of raw material can reduce within-group variance and make it easier to find differences due to the factor of interest.

Step 6: Choose the experiment structure.

For a simple comparative experiment, this is predetermined: the experiment has one factor, with levels chosen in Step 4.

There are some choices to make about the way observations are allotted to factor levels. The most advantageous arrangement is a balanced assignment: unless there is a compelling reason to do otherwise, the same number of observations should be assigned to each factor level. Balanced experiments have lower beta risk than unbalanced experiments utilizing the same resources, and because of this all the sample size tables in this text assume balanced experiments. Balanced experiments are also less adversely affected by *heteroscedastiscity,* the inequality of error term variances by factor level.

Step 7: Determine experiment risks and resource requirements.

The definitions of experimental risks for comparative experiments are the same as they are for comparisons of a single mean against a standard. Alpha risk is the probability of concluding that the factor causes a difference when it does not actually affect the response; beta risk is the probability of failing to detect an actual difference in the response caused by the factor.

Alpha risk is often fixed at 5%, and this value will be assumed.

Beta risk is often chosen to be 10% at a specific alternative (real difference) of interest. Because more than one population is involved, alternatives are a little more complex in comparative experiments than for tests against a standard.

When comparing a single population mean to a standard value, only the difference between the population average and the standard value was specified. Now differences are a vector of tube effects like: $\tau_1 = 2.5$, $\tau_2 = 3.5$, $\tau_3 = -2.0$, $\tau_4 = -4.0$.

Specifying all such differences and those of equivalent importance is impossible, so a shortcut is routinely used: specify the range of tube means (8.5 above) at which beta risk is to be controlled. Tables can then be used to determine sample size based on this range. This actually controls risk over an entire class of alternatives; the following are all in the "7.5 class":

$\tau_1 = 0.5$	$\tau_2 = 3.5$	$\tau_3 = 0.0$	$\tau_4 = -4.0$
$\tau_1 = 0.0$	$\tau_2 = 3.5$	$\tau_3 = 0.5$	$\tau_4 = -4.0$
$\tau_1 = 4.0$	$\tau_2 = -0.5$	$\tau_3 = -3.5$	$\tau_4 = 0.0$

The actual difference chosen should be large enough to be of practical significance, but not so great that such a difference would be intolerable. Since beta risk decreases as differences increase, differences bigger than the one specified are even more likely to be detected.

■ *Example 2.2: Choosing an Alternative*

A poly doping process has a target of 35 Ω and specification limits or 25 and 45 Ω. Losses from the process are known to be predicted accurately with an upside-down normal loss function with $\lambda = 8.5$. See Drain and Gough (1996) for details on this loss function. If the process mean and standard deviation are denoted by μ and σ, respectively, then the expected loss (EL) can be found as follows:

$$EL(\mu) = 1 - \frac{8.5}{\sqrt{\sigma^2 + (8.5)^2}} e^{-\left[\frac{(\mu - 35)^2}{2(\sigma^2 + (8.5)^2)}\right]}$$

The process standard deviation is typically about 0.445, so if the process mean is known the expected loss can be determined. Expected loss was computed for

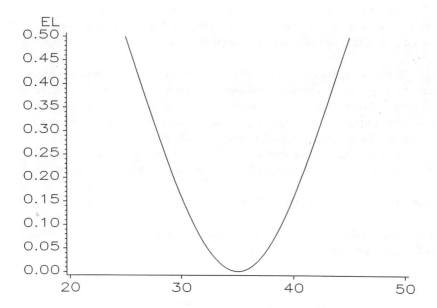

Fig. 2.3 Expected loss for a polysilicon doping process with process means ranging from the lower to the upper specification limits is plotted. The upside-down normal loss function was used to model the loss, and the scale parameter was chosen so that the loss at either specification limit would be 50%.

process means ranging from the lower to the upper specification limits; this data is plotted in Figure 2.3. Values from a portion of that graph are shown below where δ is the difference between the process mean and the process target:

δ	Expected Loss (%)
2.6	5
2.4	4
2.1	3
1.7	2
1.2	1

If a consistent loss of 3% could be tolerated, δ could be as great as 2.1 Ω; if a loss of only 1% could be tolerated, δ could not be greater than 1.2 Ω. Setting the alternative to 1.2 Ω (with $\beta = 0.10$) would ensure a 90% chance that a consistent

1% loss would not be introduced to the process as the result of a decision made on the basis of a comparative experiment. Differences that would cause larger losses would have a smaller chance of escaping detection.

One more ingredient is needed before sample size can be determined—an estimate of within-group variance. Previous experiments are an excellent source for such information, as is data from production history. Be certain that the estimate is of within-group variance and not overall variance. Using the overall variance will inflate sample size and result in unnecessarily expensive experiments.

A quick way to estimate within-tube variance from process data is to average the variances (not the standard deviations) obtained from each tube separately.

■ *Example 2.3: Estimating Within-Tube Variance*

To estimate within-tube variance from the poly doping data, compute the variance within each tube, then average these four separate estimates.

Tube	Resistivity Variance
13A	9.38
13B	11.00
13C	21.03
13D	9.54
Average	12.74

The within-tube variance estimate is 12.74. Note that variances were averaged—this method will not work if standard deviations are averaged.

Once experimental risks have been clearly specified, either Table 2.1 or the nomograph in Figure 2.4 is used to choose a preliminary sample size. Use of the nomograph will be demonstrated first because it helps build some intuitive appreciation of the relationships between sample size, number of factor levels, size of the difference to be detected (δ), and intrinsic variance (σ^2).

The nomograph was designed to find a sample size for balanced comparative experiments with 5% alpha risk and 10% beta risk against a specified alternative. Sample sizes for experiments with from two to eight factor levels are shown.

The grid on the left side of the nomograph has σ on its horizontal axis; this corresponds to within-group standard deviation (not variance). The vertical axis has δ on its axis, corresponding to a specific alternative range in factor level effects at which beta risk is to be controlled.

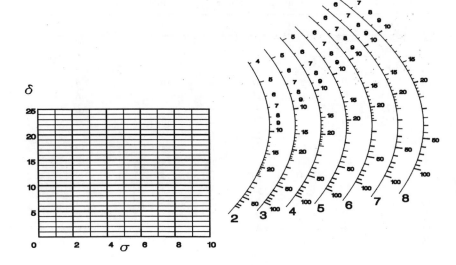

Fig. 2.4   The nomograph can be used to find a sample size for balanced comparative experiments with 5% alpha risk and 10% beta risk against a specified alternative. Sample sizes for experiments with from two to eight factor levels are included on this graph.

To use the nomograph, plot the σ-δ combination at which beta risk is to be controlled on the grid and draw a line from (0, 0) through this point. Extend the line until it intersects the curve corresponding to the number of factor levels (labeled at the bottom of the curves) in the experiment, and read the sample size.

To use the nomograph, plot the σ-δ combination on the grid and draw a line from (0,0) through this point. Extend the line until it intersects the curve corresponding to the number of factor levels (labeled at the bottom of the curves), and read the sample size.

■ *Example 2.4: Determining Sample Size with a Nomograph*

Suppose that a poly doping experiment with four levels was to be conducted with 5% alpha risk and 10% beta risk against an effect range of 6 Ω, and that within-group standard deviation was 3.57 Ω. To find sample size, draw a line from (0,0) through (3.57, 6.00) as shown in Figure 2.5. The line intersects the curve labeled 4 at about 11. This means that the experiment would require 11 independent runs on each of the four tubes, for a total of 44 runs.

Note that the sample size determined here is the number of independent observations required at each level of the factor. If an experiment has one factor with

TABLE 2.1.   Sample Size for Analysis of Variance with 5% Alpha Risk and 10% Beta Risk

					Δ				
p	*0.50*	*0.75*	*1.00*	*1.25*	*1.50*	*1.75*	*2.00*	*2.50*	*3.00*
2	85	39	22	15	11	8	7	5	4
3	103	46	27	18	13	10	8	6	5
4	115	52	30	20	14	11	9	6	5
5	125	56	32	21	15	12	9	6	5
6	133	60	34	22	16	12	10	7	5
7	141	63	36	24	17	13	10	7	5
8	147	66	38	25	18	13	11	7	6
9	154	69	39	26	18	14	11	8	6
10	160	72	41	27	19	14	11	8	6
11	165	74	42	28	20	15	12	8	6
12	171	77	44	28	20	15	12	8	6
13	176	79	45	29	21	16	12	8	6
14	181	81	46	30	21	16	13	9	6
15	185	83	47	31	22	16	13	9	6
16	190	85	48	31	22	17	13	9	7
17	194	87	49	32	23	17	13	9	7
18	198	89	51	33	23	17	14	9	7
19	202	91	52	33	24	18	14	9	7
20	206	92	52	34	24	18	14	9	7

This table is used to determine sample size for comparative experiments. The table lists sample sizes for factors with from 2 to 20 levels, for experiments with 5% alpha risk and 10% beta risk. The number of levels is the leftmost column of the table, headed by p. Specific alternatives are listed at the top of the table (the Δ-row) in units of within-group standard deviation:

$$\Delta = \frac{\delta}{\sigma}$$

If an experiment with six levels were needed to detect a difference of 12, and the within-group standard deviation was 8, then $\Delta = 1.5$, and sample size is 16.

six levels, and the sample size from the nomograph is 20, the total experiment will require 120 independent observations.

Independent observations cannot be obtained by remeasuring the same material; those measurements are obviously correlated. Independent observations cannot be obtained by measuring several different wafers in the same diffusion run—these measurements are also correlated. For the poly doping experiment, the run average (obtained from measuring three wafers in each tube) is treated as a single observation.

Sample size is influenced by three experiment design parameters: σ, δ, and the number of factor levels, *p*. Of these, δ has the greatest effect: reducing δ to 5 Ω in

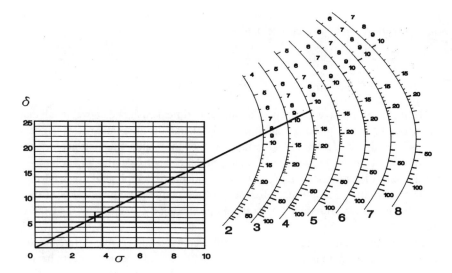

δ

Fig. 2.5 To find the sample size for an experiment with four levels, 5% alpha risk, and 10% beta risk against an effect range of 6 Ω, plot the within-group standard deviation (3.57 Ω) and the effect range on the grid. Then, draw a line from (0,0) through (3.57, 6.00) to the "4" curve. The recommended sample size is 11.

the poly doping example above would raise the sample size per tube to 16; increasing δ to 7 Ω would decrease sample size to 8.

The next greatest influence is σ. If σ increased to 5 Ω, 21 observations per tube would be required. If σ decreased to 2.4 Ω, only six runs per tube would be needed.

The number of factor levels affects the sample size in two ways: increasing the number of levels increases the sample size for each level, and this greater sample size must be multiplied by the number of levels to get the overall experiment size. For example, if two tubes were involved in the poly doping experiment, nine runs from each tube would be needed, for a total experiment size of 18. If eight tubes were involved, 14 runs would be needed from each tube, for a total experiment size of 112.

■ *Example 2.5: Determining Sample Size with a Table*

Table 2.1 can also supply sample sizes for comparative experiments. The table lists sample sizes for factors with from 2 to 20 levels, for experiments with 5% alpha risk and 10% beta risk. The number of levels is the leftmost column of the table, headed by p. Specific alternatives are listed at the top of the table (the Δ-row) in units of within-group standard deviation:

$$\Delta = \frac{\delta}{\sigma}$$

To find a sample size for the poly doping experiment designed above, recall that the within-group standard deviation was 3.57 Ω, and the alternative at which beta risk is limited was 6 Ω. Dividing the alternative by the standard deviation produces a Δ of 1.68. Read the 4 row (the factor has four levels) across to the 1.75 column— this is the choice nearest to 1.68. At least 11 independent replicates are needed.

The preliminary sample size often seems to too large to be practical or afford-able, but can often be reduced without compromising the integrity of the experiment.

First, consider choosing a larger (but still acceptable) detectable difference (δ). This quantity has the greatest effect on sample size, and many engineers have a tendency to overspecify it. A small compromise in δ can produce a much less expensive experiment.

Second, reexamine the estimate of within-group variance (σ^2) that was used. Mistaking overall variance for within-group variance is a common mistake, and one that can easily inflate sample size by an order of magnitude.

Third, consider reducing the number of factor levels. If some levels can be omitted without significantly reducing the scope of the conclusions, some reduction in sample size can be obtained.

If none of these tactics produce a sample size that still meets reasonable experiment objectives, do not proceed with the experiment; it would only waste material, and could lead to mistaken conclusions.

Step 8: Outline experiment execution.
Section 2.4 gives some direction on this topic.

Step 9: Plan data analysis and decision making.
Section 2.5 covers this topic.

2.4   EXECUTION

Before the experiment proceeds, make sure that any necessary equipment and materials are on hand, and that variables that must be controlled truly will be controlled. The people who will do the work—process wafers and make measurements—should be fully informed of the reasons for the experiment, and of the decisions that will depend on its results. They can often spot flaws in the experiment design and execution that may be invisible to an engineer.

Before and during the experiment, perform routine checks to ensure that the process is stable: an unstable process differs qualitatively from a stable process, so conclusions obtained based on data collected from it may not apply to the stable process. An unstable process also has greater variance than a stable process. This extra variance might not allow important differences to be detected.

Many of the tactics of modern experimentation are undertaken to prevent the corrupting impact of confounding variables. A *confounding variable* is some (often unsuspected) influence on a response that happens to change in unison with the experimental factor. It can make that factor appear to be more or less influential than it truly is.

■ *Example 2.6: Confounding in a Poly doping Experiment*

Suppose that a poly doping tube comparison experiment had been conducted as follows: runs for tube 13A and tube 13B were done on August 12, and runs for tubes 13C and 13D were done on August 19. Data from the experiment is summarized with boxplots in Figure 2.6. Poly resistivity seems to fall into two groups: tubes 13A and 13B are similar, and tubes 13C and 13D are similar, but the two groups are different.

Poly doping pressure was measured during normal maintenance on all of the tubes before the experiment on August 12, but on August 30 all four tubes were found to be operating at lower than their specified pressure. No pressure data is available for days between these two dates.

Doping pressure could be causing the difference in resistivity, or some difference in the tubes could be, but because tubes and pressure changed during the same time period it is impossible to tell which factor actually caused the difference. The tube factor is confounded with the pressure factor, so data from this experiment cannot be used to assess the significance of the tube factor.

Potential confounding variables should be carefully controlled during an experiment, but this can only be done if the variable is suspected to exert an influence on the response, and if it can actually be measured and controlled. Confounding variables are lurking everywhere to trap the unwary experimenter, and they are not usually as obvious as in the example above. The only known insurance against the effects of accidental confounding is randomization.

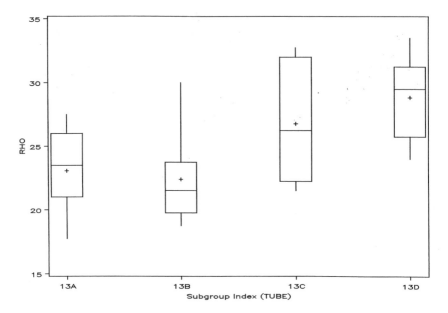

Fig. 2.6   Data from a poly doping tube comparison experiment is summarized with boxplots. Each boxplot corresponds to one poly doping tube.

For a simple comparative experiment, *randomization* refers to the random assignment of experimental units to factor levels. A randomized experiment is not a careless experiment: randomization does not mean that factor levels are chosen at random, or that variables that should be controlled are allowed to range freely.

■ *Example 2.7: Randomization*

An experiment is to be done to evaluate the effect of poly doping tubes on poly silicon resistivity—11 replicates will be done.

Before the experiment, the names of the four poly doping tubes were each written on 11 pieces of paper, so there was one piece of paper for every one of the 44 runs in the upcoming experiment. The pieces of paper were mixed up in a hat, and drawn out at random with the following result:

13D, 13D, 13B, 13B, 13A, 13C, 13B, 13C, 13A, 13D, 13C,
13D, 13B, 13C, 13A, 13C, 13B, 13A, 13C, 13D, 13C, 13D,
13D, 13A, 13A, 13B, 13D, 13A, 13A, 13D, 13B, 13C, 13D,
13B, 13A, 13B, 13D, 13A, 13C, 13B, 13B, 13C, 13A, 13C.

Note that every poly doping run was individually assigned to a tube; randomization does not allow all settings of a factor to be run together, so all the runs on tube 13A do not occur in one contiguous span of time. Some consecutive runs are on the same tube—13D and 13D at the start of the experiment for example; this coincidence is an acceptable outcome of random assignment.

Randomization is not an absolute guarantee that confounding variables will not affect responses, but it does lessen the probability that their influence will lead to erroneous conclusions. Confounding variables still increase response variance even in a randomized experiment, so they should be eliminated or controlled whenever they are known.

Randomization does have a cost, and in some experiments small compromises to total randomization may be necessary for economic reasons. For example, if two subjects are to be used to test new bunnysuits for particulation and only one measurement chamber is available, it seems sensible to have the subjects alternate tests throughout the experiment. (*Bunnysuit* is the term used to describe the white coveralls which are worn in a cleanroom, probably because someone thought they made people look like big white rabbits.) In this way subject A can be changing into the next bunnysuit while subject B is in the test chamber, and the entire suite of tests can be completed as quickly as possible.

This seems like an excusable restriction: measurements on both individuals are spaced over time, which in this experiment is often correlated with most known confounding factors like subject fatigue, test chamber temperature changes, and measurement system drift. A systematic confounding variable that interfered with alternate tests could completely invalidate experimental results, but such variables seem unlikely in this context.

■ *Example 2.8: Restricted Randomization*

Consider an experiment in which the temperature of an oven was a factor. Three temperature levels were to be tested and four replicates at each temperature were needed. The oven is quick to heat to a given temperature but slow to cool down afterward, so the experiment could proceed quicker if temperature decreases could be avoided.

One complete randomization scheme was initially chosen that would have required a total of six temperature decreases:

| Low | High | Low | Medium | High | Medium |
| High | High | Low | Medium | Medium | Low |

This scheme was adapted to avoid some of the temperature decreases. A partial randomization scheme put these same runs into four groups, and then sorted by temperature within group:

	Order in Group		
Group	*1*	*2*	*3*
1	Low	Low	High
2	Med	Med	High
3	Low	High	High
4	Low	Med	Med

Now only three temperature decreases must occur. The temperature grouping saved enough time for all runs to be completed in a single day, but still allowed for a substantial degree of randomization.

Any systematic assignment of treatments to experimental units must be carefully scrutinized. An experiment investigating the effect of two resist cleaning methods on yield was conducted by assigning odd numbered wafers to the first cleaning method, and even numbered wafers to the second. Unfortunately, wafers were always processed in numerical order, and wafer 1 invariably had lower yield because it was situated nearest the back of a (dirty) wafer carrier. This made it appear that the first cleaning method was inferior to the second, when it really was not.

Be skeptical of randomization restrictions and talk over any planned restrictions with a statistician or engineers familiar with the process. Problems due to faulty randomization are easy to prevent, but data from flawed experiments may be impossible to interpret usefully.

2.5   ANALYSIS AND INTERPRETATION

Section 2.5.1 shows how to estimate effects (the τ_i in the model), both with point estimation and interval-based means comparisons. Section 2.5.2 shows how to apply ANOVA to comparative experiments to arrive at a decision regarding the significance of a factor effect.

The assumptions underlying the model are critical—they must be examined before conclusions can be accepted and reported. Section 2.5.3 is devoted to assumption checking.

Section 2.5.4 gives some advice on reporting experiment results in a manner that can be understood by other experimenters and management.

2.5.1 Effects Estimates and Means Comparisons

Once an experiment has been completed, the average response for each factor level can be found, as well as the overall experiment average. The difference between each group mean and the overall mean is an estimate of the effect of the factor at that level.

■ *Example 2.9: Effects Estimates*

The poly doping experiment was run in randomized order as outlined in Example 2.8 above, with the following results:

13A	13B	13C	13D
25.00	20.50	31.00	25.50
24.00	16.00	32.00	30.25
22.00	14.50	31.00	30.00
22.00	21.00	32.75	25.00
19.75	21.25	27.75	25.00
19.00	17.00	29.75	24.75
21.75	19.75	25.50	30.50
18.25	19.75	26.50	22.75
17.25	18.00	31.25	29.75
23.50	19.25	25.50	26.75
22.25	16.25	30.75	29.75

The grand mean (24.13) was subtracted from each of the group means to arrive at point estimates for the effects:

Tube	Mean	Effect
13A	21.34	−2.79
13B	18.48	−5.65
13C	29.43	5.30
13D	27.27	3.14

Tubes 13A and 13B appear to decrease poly resistivity, while tubes 13C and 13D seem to increase it.

These estimates suffer from the shortcomings of any point estimate: they give neither a measure of their precision, nor an assessment of their probability of correctness.

Confidence interval estimates for the difference between any pair of factor level means can be estimated in a variety of ways; see Miller (1981) for a complete discussion of the topic. In this text, Tukey's procedure has been chosen for general use. The Tukey means comparison procedure seems to work well and is widely available in software packages. The computations underlying the method are not shown here; Neter et al. (1985) give details.

Tukey's procedure produces simultaneous confidence intervals for differences in factor level effects. Simultaneous confidence intervals can be used together to make conclusions about the relationship between any of the estimated quantities, and the overall confidence level is maintained regardless of the number of intervals involved. Individual (nonsimultaneous) confidence intervals apply to only one interval at a time, so if several intervals are used together, the overall confidence of the conclusion they support is eroded.

For example, if the Tukey procedure is applied in an experiment with four factor levels and overall 95% confidence, all six intervals can be used at once to make conclusions about factor effects, and with 95% confidence. If a nonsimultaneous procedure were used, these six comparisons could be used simultaneously with only 70% confidence because the risk from each estimation adds to the overall risk.

The issue of simultaneous testing is a deeper subject than it may seem from this discussion; the topic is treated thoroughly in Neter et al. (1985). Westfall and Young (1993) give a contemporary approach to the issue of multiple testing in general.

■ *Example 2.10: Tukey Multiple Comparison Procedure*

One form of the Tukey multiple comparison procedure gives confidence intervals for the difference between group means. SAS output for this is shown below for the poly resistivity data. Note that the confidence level for the set of intervals is 95%.

According to this output, the effects of tubes 13C and 13D do not differ from one another: the confidence interval for their difference is (–0.767, 5.085), and this interval contains zero. Tubes 13A and 13B also have similar effects, but the two pairs of tubes (13A–13B and 13C–13D) do have significantly different effects.

An alternate form of the output groups factor levels according to whether their means differ or not; levels that do not differ are connected with a line. This is also a simultaneous testing procedure; the overall alpha risk is 5%.

This printout below indicates that tubes 13C and 13D are in one group, and tubes 13A and 13B are in a different group.

General Linear Models Procedure: Tukey's Studentized Range (HSD) Test for Variable: Rho*

Tube Comparison	Simultaneous Lower Confidence Limit	Difference Between Means	Simultaneous Upper Confidence Limit	
13C–13D	−0.767	2.159	5.085	
13C–13A	5.165	8.091	11.017	***
13C–13B	8.029	10.955	13.881	***
13D–13C	−5.085	−2.159	0.767	
13D–13A	3.006	5.932	8.858	***
13D–13B	5.869	8.795	11.721	***
13A–13C	−11.017	−8.091	−5.165	***
13A–13D	−8.858	−5.932	−3.006	***
13A–13B	−0.062	2.864	5.790	
13B–13C	−13.881	−10.955	−8.029	***
13B–13D	−11.721	−8.795	−5.869	***
13B–13A	−5.790	−2.864	0.062	

Abbreviations: dF, degrees of freedom; MSE, mean square due to error.
*This test controls the Type I experimentwise error rate. Alpha risk = 0.05; Confidence = 0.95; df = 40; MSE = 6.553977; critical value of studentized range = 3.791; minimum significant difference = 2.926; comparisons significant at the 0.05 level are indicated by ***.

Means with the same letter are not significantly different.

Tukey Grouping	Mean	n	Tube
A	29.432	11	13C
A			
A	27.273	11	13D
B	21.341	11	13A
B			
B	18.477	11	13B

Means comparison procedures clarify factor effects by giving details on which levels of the factor seem to be causing the overall effect. However, if there is no factor effect, these comparisons are unnecessary. ANOVA is a shortcut to esti-mating the overall factor effect, and it is usually the first test done after assumptions have been verified.

2.5.2   Analysis of Variance

ANOVA provides a hypothesis test that generates a definitive decision regarding the significance of a factor effect on a response.

The hypotheses tested are:

$$H_0 : \tau_1 = \tau_2 = \ldots = \tau_p = 0$$
$$H_A : \tau_i \neq 0 \quad \text{for some} \quad i \in \{1, 2, \ldots, p\}$$

If H_0 is rejected, the factor is said to have a significant effect on the response. If H_0 is not rejected, then the effect was not statistically significant—it could be zero, or it could simply be too small to detect with that particular experiment.

■ *Example 2.11: Hypotheses for a Poly doping Experiment*

For the poly doping experiment, four levels (tubes) of the factor are of interest, so the hypotheses to be tested are:

$$H_0 : \tau_1 = \tau_2 = \tau_3 = \tau_4 = 0$$
$$H_A : \tau_i \neq 0 \quad \text{for some} \quad i \in \{1, 2, 3, 4\}$$

where τ_i is the effect of the ith tube.

The basis of ANOVA is not complicated—it is a simple comparison of two es-timates of variance. The principle is the same one used to justify the analysis of linear regression data. Computer programs can do all the computations required for ANOVA, so an understanding of its inner workings is not necessary to suc-cessfully use ANOVA.

The first step is to estimate the variance due to the factor; this is usually called the *mean square due to the treatment,* or MST. MST is found by computing the sample variance of the group means, and then multiplying by the number of ob-servations in each subgroup, n. For the poly doping data above, the variance of the four tube means (21.34, 18.48, 29.43, 27.27) is 25.91. Multiplying by 11 yields an MST of 285.0.

Next, estimate within group variance directly by averaging the four within-group variances. This quantity is called *mean square due to error,* denoted by MSE:

Tube	Resistivity Variance
13A	6.13
13B	5.13
13C	7.00
13D	7.96
Mean	6.55

The estimate of MSE for the poly doping data is 6.55.

Once these two variance estimates are obtained, take their ratio to obtain the sample F statistic, or F_0:

$$F_0 = \frac{\text{MST}}{\text{MSE}}$$

The sample F should be compared with $F_{p-1, p(n-1), 0.95}$. If the sample F is greater than this percentile, reject H_0.

■ *Example 2.12: F-Test*

As seen above, MST = 285.0 and MSE = 6.55; hence, the sample F ratio is 43.51. Because 11 runs were made on each of four tubes, $p = 4$ and $n = 11$ so the degrees of freedom for the corresponding F-percentile are $v_1 = 4 - 1 = 3$ and $v_2 = 4(10) = 40$. According to Table A.1, the 95th percentile of the $F_{3,40}$ distribution is 2.84." Because the observed ratio is larger than this critical value, H_0 will be rejected: the poly doping tube factor has a significant effect on resistivity.

Statistical packages do exactly this test when they perform analysis of variance, although the appearance of the test may differ.

■ *Example 2.13: ANOVA with SAS*

An edited SAS output for the poly doping data is shown below. The F Value shown on the printout is the same as the sample F computed above by hand. The actual decision process is simple: for an $\alpha = 0.05$ test, if the probability of observing this F-ratio under the null hypothesis ("Pr > F") is less than 0.05, then H_0 should be rejected.

General Linear Models Procedure

Class level information

Class	Levels	Values
Tube	4	13A 13B 13C 13D

No. of observations in data set: 44

Dependent variable   rho

Source	df	Sum of squares	F Value	Pr > F
Model	3	854.90198864	43.48	0.0001
Error	40	262.15909091		
Corrected total	43	1117.06107955		

R-square	CV	Rho mean
0.765314	10.60921	24.13068182

Source	df	Type I SS	F value	Pr > F
Tube	3	854.90198864	43.48	0.0001

Abbreviations: CV, Coefficient of Variation; SS, sum of squares.

In this case, the Pr > F value is 0.0001, meaning that the observed values lies at the 99.99th percentile of the $F_{3,40}$ distribution. This is greater than the 95th percentile, so H_0 will be rejected.

MST and MSE can be found from the model sum of squares and the error sum of squares by dividing by their respective degrees of freedom (df):

$$MST = 854.90 \div 3 = 284.97$$
$$MSE = 262.16 \div 40 = 6.55$$

These values agree (up to roundoff error) with the calculations made by hand.

The square root of MSE is a useful quantity because it can be used as an estimate of σ when planning future experiments.

The hypothesis test has only two outcomes: it either indicates that the factor has a (statistically) significant influence on the response, or it does not. Either of these outcomes can occur erroneously if the ANOVA assumptions are violated, so the actions outlined in Section 2.5.3 are critical to ensuring the validity of conclusions.

If the factor effect is significant, level-specific effects should be estimated and means comparison methods used to see which levels differ from the overall mean, and how. In some cases, the magnitude of the factor effect may not be of practical significance—although statistically significant, it may be so small that no reasonable person would act to change the process to eliminate the effect.

If the factor effect does not appear to be significant and the model assumptions have been verified, then a beta risk curve can be used to quantify the probability that a factor effect did exist but was not detected. This post hoc examination of beta risk should be unnecessary if the experiment was designed to find differences of practical importance, but the beta risk curve is still useful when explaining experimental risks and results to others. Beta risk curves are essential in the evaluation of experiments where design goals were not documented.

Beta risk curves for common sample sizes and numbers of levels from two through nine can be seen in Figures 2.7–2.14. Others can be obtained from statistical software packages.

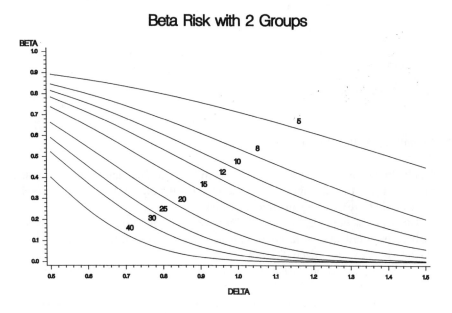

Fig. 2.7   This set of beta risk curves applies to experiments with two levels and 5% alpha risk. Beta risk is plotted on the vertical axis; the horizontal axis is a range of standardized differences (actual difference divided by within-group standard deviation). Each curve is identified by sample size—the number of replicates per factor level.

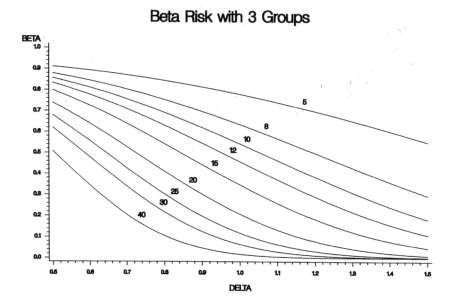

Fig. 2.8   This set of beta risk curves applies to experiments with three levels and 5% alpha risk. Beta risk is plotted on the vertical axis; the horizontal axis is a range of standardized differences (actual difference divided by within-group standard deviation). Each curve is identified by sample size—the number of replicates per factor level.

■ *Example 2.14: Beta Risk Curve*

A poly doping experiment with four levels (tubes) was conducted, but no significant effect was found. Engineers wanted to know the probability that the experiment would have detected an effect of 1.80 Ω.

The beta risk curve in Figure 2.9 can answer this question. This curve shows the beta risk (vertical axis) faced for a range of standardized differences on the horizontal axis. A standardized difference is the actual difference divided by within-group standard deviation. Risk differs according to the number of replicates at each level, so sample size is denoted on each curve.

The within-tube standard deviation is estimated by the square root of the within-tube variance—the square root of MSE. MSE for this experiment was 6.55, so σ is estimated to be 2.56. Dividing the alternative (1.80) by this value gives a standardized difference of 0.70. The number on the beta risk axis corresponding to this difference and a sample size of 11 (interpolated) is about 73%.

Beta Risk with 4 Groups

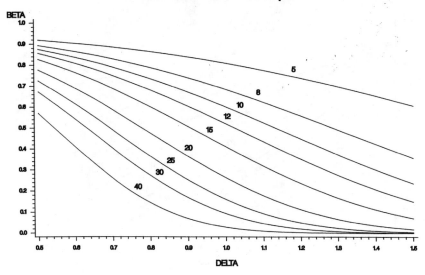

Fig. 2.9   This set of beta risk curves applies to experiments with four levels and 5% alpha risk. Beta risk is plotted on the vertical axis; the horizontal axis is a range of standardized differences (actual difference divided by within-group standard deviation). Each curve is identified by sample size—the number of replicates per factor level.

2.5.3   Checking Assumptions

The calculations and statistical tests used in ANOVA are based on some assumptions that must be verified; if the assumptions are violated, ANOVA can give completely deceptive results.

The most important assumption relates to the distribution of the error terms of the model, which is: ε_{ij} is a random error for the jth replicate at the ith level of the factor; these errors are assumed to be independent, and are normally distributed with mean zero and constant variance σ.

Error terms cannot be directly observed, but they can be estimated by residuals, e_{ij}, which are found by subtracting group means from observed responses:

$$e_{ij} = y_{ij} - \overline{y}_i$$

for $i = 1 \ldots p$, where p is the number of levels of the factor.

Beta Risk with 5 Groups

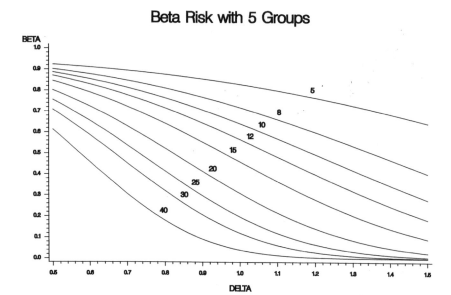

Fig. 2.10   This set of beta risk curves applies to experiments with five levels and 5% alpha risk. Beta risk is plotted on the vertical axis; the horizontal axis is a range of standardized differences (actual difference divided by within-group standard deviation). Each curve is identified by sample size—the number of replicates per factor level.

Beta Risk with 6 Groups

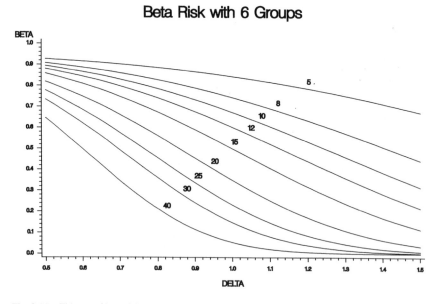

Fig. 2.11   This set of beta risk curves applies to experiments with six levels and 5% alpha risk. Beta risk is plotted on the vertical axis; the horizontal axis is a range of standardized differences (actual difference divided by within-group standard deviation). Each curve is identified by sample size—the number of replicates per factor level.

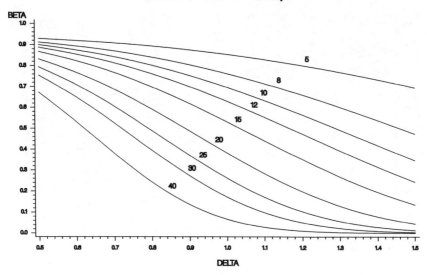

Fig. 2.12   This set of beta risk curves applies to experiments with seven levels and 5% alpha risk. Beta risk is plotted on the vertical axis; the horizontal axis is a range of standardized differences (actual difference divided by within-group standard deviation). Each curve is identified by sample size—the number of replicates per factor level.

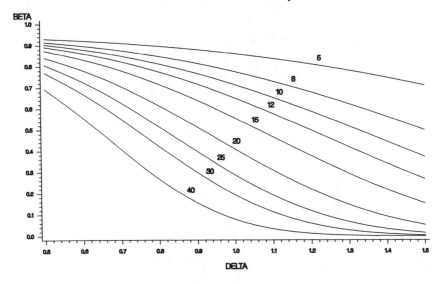

Fig. 2.13   This set of beta risk curves applies to experiments with eight levels and 5% alpha risk. Beta risk is plotted on the vertical axis; the horizontal axis is a range of standardized differences (actual difference divided by within-group standard deviation). Each curve is identified by sample size—the number of replicates per factor level.

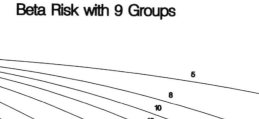

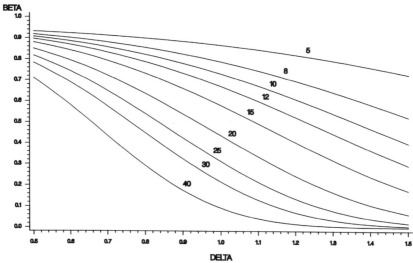

Fig. 2.14   This set of beta risk curves applies to experiments with nine levels and 5% alpha risk. Beta risk is plotted on the vertical axis; the horizontal axis is a range of standardized differences (actual difference divided by within-group standard deviation). Each curve is identified by sample size—the number of replicates per factor level.

RES_RHO

Fig. 2.15   A plot of residuals in time order should not display any patterns indicating that residuals influence one another. Residuals in this plot do appear to be independent.

■ *Example 2.15: Calculating Residuals*

Residuals for the first five observations from the poly doping experiment are shown below. The "tube mean" is the observed average resistivity for that tube; "observed resistivity" is the measured resistivity for that particular run.

Tube	Observed Resistivity	Tube Mean	Residual
13D	25.50	27.27	–1.77
13D	30.25	27.27	2.98
13B	20.50	18.48	2.02
13B	16.00	18.48	–2.48
13A	25.00	21.34	3.66

Most assumptions relating to the model can be checked with a few quick plots of residuals. A plot of residuals in time order (shown in Figure 2.15 for the poly doping data) should not display any patterns indicating that residuals influence one another.

The shift apparent in the residual plot in Figure 2.16 was caused by an unstable process; the overall process mean changed during the experiment, so the relationship of individual measurements to their respective tube means also changed.

The "triplet" correlation pattern in Figure 2.17 is due to a misinterpretation of the raw data. Three wafers were measured in each run, and rather than recording their mean, the individual values were treated as independent replicates. Wafers within a tube during the same run are highly correlated, so the assumption of error independence is violated. This type of counterfeit replication is a common mistake, and it is likely to cause false positives in the hypothesis test because MSE is artificially deflated.

A histogram of residuals (Figure 2.18) can verify their normality. The most common source of nonnormality is outliers due to measurement or transcription

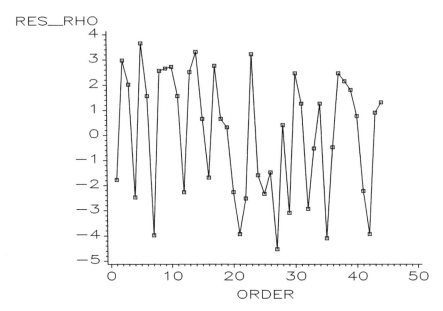

Fig. 2.16   This plot of residuals in time order shows a shift caused by a process that was unstable during the course of an experiment.

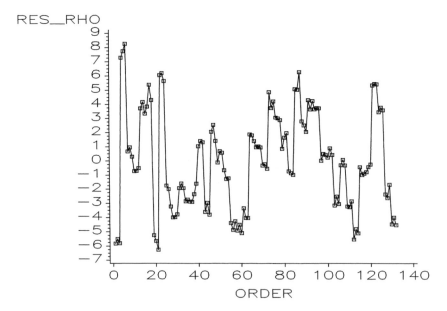

Fig. 2.17   This plot of residuals in time order shows a systematic pattern of correlation: every "triplet" of points seems to be highly correlated.

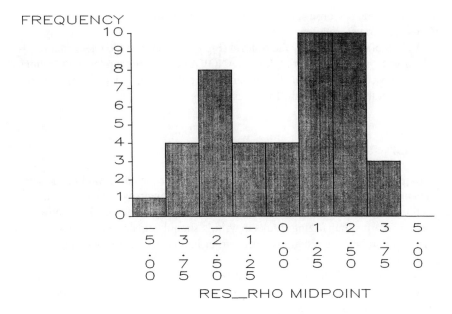

Fig. 2.18   This histogram of residuals seems to verify their normality.

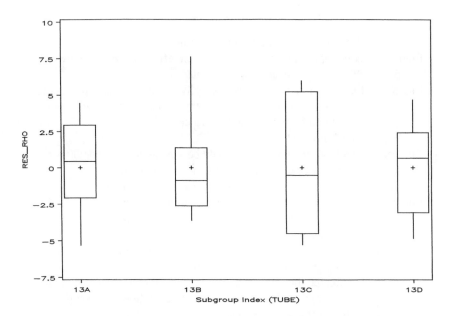

Fig. 2.19   These boxplots of residuals by factor level show that residual variance is reasonably uniform over all levels of the factor.

mistakes. Some responses are naturally not normally distributed, and may require the use of a transformation.

A boxplot of residuals by factor level (Figure 2.19) is used to check that residual variance is uniform over all levels. ANOVA can tolerate moderate deviations from homoscedastiscity (equality of variance), but excessive departures will seriously affect the validity of the hypothesis test.

2.5.4   Reporting Results

Experiments should be well documented, both for the edification of management and other engineers, and so that future experimenters will have estimates of within-group variance and other important information. The checklist shown in Figure 2.20 for summarizing results in writing should help organize a final report; it closely parallels the blueprint given earlier in this chapter for planning the experiment.

The language of hypothesis testing is somewhat counter-intuitive; results seem to be stated in a double-negative form which can be quite confusing. Three sample dialogues are given below to show how results can be clearly presented.

Dialogue 1: The effect is significant.

Boss: What was the result of this experiment?

Engineer: Poly doping tubes have a significant effect on poly resistivity.

Boss: What do you mean by significant?

Engineer: I mean that the effect is statistically significant—an apparent effect as great as the one I saw in this experiment would be extremely unlikely to occur by chance.

Boss: How unlikely?

Engineer: An effect this large would only be seen about 1 in 10,000 times, if there were really no effect there to be observed.

Boss: What do you mean exactly, that there is an effect?

Engineer: Some of the tubes produce poly with resistivity higher than the overall average—13C and 13D. Tubes 13A and 13B produce poly with resistivity lower than the overall average.

Boss: All right, so the effect is statistically significant; how do you know it is big enough to make any difference?

Engineer: I estimated the size of the effect; the difference between the highest and lowest tubes' resistivity is 11 Ω. I think that is a practically significant difference.

Boss: How do you know that you just didn't get unlucky? Maybe tubes 13A and 13B ran early in the week when the preclean station was acting up.

Simple Comparative Experiment Summary

☐ What factor will be tested in this experiment?
☐ What is the scope of the experiment: About what population or equipment and processing conditions will inferences be made?
☐ Has this factor previously been found to be significant?
☐ What conditions must be carefully controlled for this experiment to succeed?
☐ Can a reliable estimate of within-group variance be obtained?
☐ What is the indicative response - the one which originally motivated the experiment?
☐ Are there other responses, either indicative or economic?
☐ How will each of these responses be measured?
☐ Does measurement equipment has sufficient measurement capability to detect important differences in any of the chosen responses?
☐ Exactly which factor levels will be used?
☐ Are each of these levels necessary?
☐ Is this set of levels sufficient to satisfy the scope of the experiment?
☐ Which other variables will be controlled or measured before and during this experiment? Include any environmental, process, or human sources of variance which would obscure the effect of the factor.
☐ Is this a balanced experiment?
☐ What alpha risk has been chosen? (Usually 5%).
☐ What specific alternative (range of factor level effects) was used to determine sample size?
☐ What beta risk with respect to this specific alternative has been chosen? (Usually fixed at 10%)
☐ What number of replicates at each factor level are needed?
☐ Were all observations taken as planned? What data was actually collected?
☐ Were all routine checks done as directed?
☐ Were variables which should have been controlled actually controlled?
☐ Did anything occur which indicated that the process was unstable during the experiment?
☐ Explain any restrictions on randomization. Could any of these have compromised the validity of the results?
☐ Do residuals appear to be normally distributed and uncorrelated.
☐ Is residual variance about the same for each factor level?
☐ Is there anything which might compromise the validity of these results?
☐ What was the result of the ANOVA hypothesis test?
☐ If the factor effect was significant, provide estimates of effects. Are any of these effects practically significant?
☐ If the factor effect was not significant, use the estimate of within-group variance from the ANOVA table to construct a beta risk curve.

Fig. 2.20   A checklist for experiment documentation which helps to summarize results a final report.

Engineer: I randomized the order of the runs, so problems like that couldn't affect my conclusions.

Boss: Well, it sounds like you found a real difference. What are you going to do about it?

Engineer: I'll examine the tubes to see what 13A and 13B have in common, and what 13C and 13D have in common, and how these two groups of tubes differ. I should be able to find some equipment disparity which would explain their differing results.

Boss: Good job. Let me know what you find and I'll make sure the other diffusion engineers check their equipment for similar problems.

Dialogue 2: The effect is not significant

Boss: What was the result of this experiment?

Engineer: Tubes do not appear to have a significant effect on the resistivity of the poly they produce.

Boss: What do you mean, "do not appear to have a significant effect"? They have an effect, or they don't.

Engineer: You're absolutely right—the tubes either have an effect or they don't. This experiment did not detect any effect.

Boss: Maybe you just did the experiment wrong.

Engineer: I designed the experiment to detect an effect of 6 Ω 90% of the time. The fact that no statistically significant difference between the tubes was found leads me to believe that any difference that does exist is certainly less than 6 Ω.

Boss: 6 Ω is a pretty big difference; even 3 Ω would be a concern to me.

Engineer: According to the beta risk curve I've got here, I still had a 50% chance of detecting a 3 Ω difference.

Boss: Half a chance doesn't seem like very good odds to me.

Engineer: To detect a difference of 3 Ω 90% of the time would have required 27 runs on each tube. I thought that would have been an excessive amount of material to commit to this experiment.

Boss: Good decision—for that kind of money we can live with a 3 Ω offset.

Dialogue 3: Results were inconclusive.

Boss: What was the result of this experiment?

Engineer: I can't really make any conclusions based on this experiment.

Boss: Why not? We committed 44 poly runs to the experiment; we should be able to find out something.

Engineer: Unfortunately, the preclean station was contaminated sometime
after the experiment started, and it probably caused the steady increase in resistivity we observed over time.

Boss: That's unfortunate, but it shouldn't be a problem if you randomized the order of the runs.

Engineer: I thought I could get the experiment done faster if I did all of the
runs for each tube at once: all the runs for 13A were done first,
then those for 13B, then 13C, then 13D. I really can't tell if the
change in resistivity is because of the tubes, or because of the preclean station contamination.

Boss: We spent 44 runs' worth of material and lots of our operators'
time on this experiment. Can't we get any useful information out
of all that data?

Engineer: I can't see any way to salvage this experiment.

2.6   COMPARATIVE EXPERIMENTS WITH
UNUSUAL ASSUMPTIONS

Not every type of data satisfies the usual ANOVA assumptions. Particle data is
one very important source of information that is rarely normally distributed; Section 2.6.1 shows an analysis method that will produce valid results for data that
has a Poisson distribution. For those cases where the data distribution is continuous but not normal, a general nonparametric (distribution-free) method of analysis is demonstrated in Section 8.6.2.

2.6.1   Comparison of Poisson Means

Suppose that a response in a comparative experiment was the observed number of a particular type of rare visual defect. Count data like this often has no suitable normalizing transformation, so it cannot be made to fit the standard ANOVA
assumptions. Many types of count data can be successfully modeled with the Poisson distribution.

If p levels of the factor are involved, the hypotheses of interest are:

$$H_0: \lambda_1 = \lambda_2 = \ldots = \lambda_p$$
$$H_1: \lambda_i \neq \lambda_j \quad \text{for some} \quad i \neq j$$

where λ_i is the defect density for the ith level.

If $p = 2$, the test can be constructed using the binomial distribution. Under H_0
the probability of a defect being observed is exactly proportional to the area inspected. Defects are hitting both parts of the experiment randomly, so the number

of defects depends only on their opportunity of being seen. Hence, if A_1 is the area inspected in the first population, and A_2 is the area inspected in the second population, then under H_0 the number of defects observed in the first population should have a $b(n,p)$ (binomial) distribution with:

$$p = \frac{A_1}{A_1 + A_2}$$

and n equal to the total number of defects observed in both samples.
    In this simple case, the test reduces to:

$$H_0 : p = \frac{A_1}{A_1 + A_2}$$

$$H_1 : p \neq \frac{A_1}{A_1 + A_2}$$

These hypotheses can be tested with the exact binomial test.

■ *Example 2.16: Comparison of Defect Density in Two Populations*

A total of 20 defects were observed in an experiment on two poly doping tubes in which a total of 20.0 cm^2 were inspected. Six defects were observed in 5 cm^2 inspected from the first tube, and 14 were observed in 15 cm^2 inspected from the second. Given the relative areas inspected, the first tube would be expected to have 25% of the defects:

$$p = \frac{5}{5 + 15}$$

The probability of observing six or more defects given this proportion of success on each of 20 trials is:

$$\sum_{i=6}^{20} \binom{20}{i} (0.25)^i (0.75)^{20-i}$$

The probability is 0.3828, which is larger than 0.05: there is insufficient evidence to reject H_0.

    If more than two factor levels are involved the binomial distribution no longer applies, but the same principle holds: under H_0, the number of defects observed at each level should be proportional to the area inspected at that level. If A_i is the area inspected at level i, and a total of N defects are observed, then the expected number of defects at each level, E_i, would be given by:

$$E_i = \frac{NA_i}{\sum\limits_{j=1}^{p} A_j}$$

The difference between the number of defects actually observed (n_i), and the number that would be expected under H_0, (E_i), is a measure of the unequal distribution of defects among factor levels. This is summarized by the following statistic:

$$\chi^2_{\text{sample}} = \sum_{i=1}^{p} \frac{(n_i - E_i)^2}{E_i}$$

which, under H_0, has an approximately χ^2 distribution with $p-1$ degrees of freedom (Nelson, 1987). The hypothesis test consists of computing the sample chi-squared statistic and comparing it with the 95th percentile of the chi-squared distribution with $p-1$ degrees of freedom; if the sample statistic is greater than the distribution percentile, then H_0 should be rejected.

■ *Example 2.17: Comparison of Defect Density from Several Populations*

A total of 66 defects was observed in an experiment on four poly doping tubes in which a total of 34.75 cm^2 was inspected. Since four levels were involved in the experiment, the sample chi-squared will be compared with the 95th percentile of the χ^2_3 distribution. That percentile can be found (in tables) to be 7.81.

The sample chi-squared statistic is computed in the table below.

Tube	Defects Observed	Area Inspected	Expected Number of Defects	Contribution to Sample Chi-Squared
1	12	8.00	15.19	0.6715
2	14	6.25	11.87	0.3820
3	25	11.00	20.89	0.8077
4	15	9.50	18.04	0.5133
		Sample chi-squared		2.375

The sample chi-squared is 2.375, which is smaller than the table percentile (7.81); hence, there is insufficient evidence to reject H_0.

2.6.2   Nonparametric Comparisons

The Kruskal-Wallis test is used to simultaneously test the equality of a set of population medians. The test can be applied to the hypotheses:

$$H_0: \eta_1 = \eta_2 = \ldots = \eta_p$$
$$H_1: \eta_i \neq \eta_j \quad \text{for some} \quad i \neq j$$

where η_i is the median of the ith level.

The test requires only that the sampled populations are continuous and symmetric about their medians, so many populations failing to meet the ANOVA assumptions because of nonnormality can be tested with the Kruskal-Wallis test.

The test is computationally intensive and details are not given here; the interested reader can consult Hollander and Wolfe (1973) for further information. Many statistical software packages can perform the test, as is demonstrated in Example 2.18 below.

■ *Example 2.18: The Kruskal-Wallis Test*

The poly doping experiment data was analyzed with a Kruskal-Wallis test; results are shown below.

Wilcoxon Scores (Rank Sums) for Variable Rho Classified by Variable Tube*

Tube	n	Sum of Scores	Expected Under H_0	Std. Dev. Under H_0	Mean Score
13D	11	331.0	247.500000	36.8717135	30.0909091
13B	11	89.0	247.500000	36.8717135	8.0909091
13A	11	169.0	247.500000	36.8717135	15.3636364
13C	11	401.0	247.500000	36.8717135	36.4545455

*Average Scores were used for Ties. Kruskal-Wallis test (chi-square Approximation): CHISQ = 34.103; df = 3; Prob > CHISQ = 0.0001

The test detected a significant effect, as indicated by the very small Prob > CHISQ= value. For a 5% alpha risk test, any value less than 0.05 would indicate statistical significance.

2.7 SUMMARY

Simple comparative experiments are a very common type of experiment in industry, so understanding their design, performance, and analysis is essential. Analysis of variance is the main tool used to analyze these experiments, and it is also the basis for decision making in more complex experiment designs.

Some principles underlying effective and efficient experiment design were presented in this chapter will be used throughout the remainder of the book: the importance of process stability to successful experimentation, the use of randomization to prevent confounding, the need to critically examine experimental data and check assumptions, and the necessity of establishing clear goals.

CHAPTER 2 PROBLEMS

1. A kicker for a college football team desires to test three competing brands of kicking shoes to see which gives him the longest distance kicks. From past performance, he knows that his average kick travels 66 yards, with a standard deviation of 5 yards. He is interested in detecting a difference of 2.5 yards from shoe to shoe.
   (a) How many kicks must he make with each shoe to detect this difference with 5% alpha risk and 10% beta risk?
   (b) Do you think complete randomization would be necessary for this experiment? Understanding that changing shoes takes 2 minutes, how would you set up this experiment?

2. A competitive cyclist is interested in improving his performance by replacing the hardened steel axle in the bottom bracket of his bicycle with either a $500 hollowed-out titanium axle, or a $1200 ceramic axle. He plans to run an experiment in which he rides the same course multiple times using the different axles. From many previous rides, he knows that his average time to ride the course is 23 minutes, 17 seconds, with a standard deviation of 14 seconds. Since he will not buy the expensive axles if they give slower times, he will use a one-sided test.
   (a) Determine the sample size he must use to determine a 4 second difference with alpha = beta = 0.05.
   (b) Do you think this is a reasonable sample size? If the sample size is too large, but a 4-second difference is actually important, how should the cyclist proceed?

★ 3.   You find yourself in a difficult position at your new job with HSC (Hades Semi-conductor Corporation): your boss wants you to design an experiment which, "Has absolutely no chance of indicating false effects, and will detect any real effect—no matter how small." Prepare a 1-minute discourse to educate this person on the subject of experimental risks.

4.   For a timed etch process, it is critical that etch rates be consistent among all etchers in an equipment group. An engineer responsible for a group of three of these etchers sets up an experiment in which etch rate will be measured four times on each machine. He records the following data, given in Å/s:

Etcher A	Etcher B	Etcher C
5.72	6.43	5.84
4.22	6.08	7.26
4.76	5.82	6.96
5.35	6.32	6.86

(a) What is the response variable, and what is the experimental factor.
(b) With alpha = 0.05 and beta = 0.10, how small a difference should this design be able to detect?
(c) Compute a point estimate for the effect of each etcher.
(d) Using ANOVA, do you find a difference between the three etchers at a 0.05 level of significance?
(e) Using information from this analysis, determine the sample size necessary to detect a difference half the size of that found in part (b).
(f) Compute and plot residuals for this experiment. Are they normally distributed?

★ 5.   A metrology engineer is interested in whether the four scanning electron microscopes for which she is responsible are consistent in how many particles they add to wafers that they measure. She reads five particle wafers on each system, getting the following counts:

SEM1	SEM2	SEM3	SEM4
41	35	53	34
37	44	63	35
33	30	36	44
41	34	44	43
30	53	29	27

(*a*) Since this is discrete data, so you think ANOVA is a valid analytical technique? Would knowing that the data came from a Poisson distribution help?

(*b*) If you do use ANOVA, do you find that there is a significant difference between any of the SEMs?

(*c*) Assume that the same wafer area was inspected for each SEM, and use the methods in Section 2.6.1 to test for a difference in particle density.

6. An automobile manufacturer wants to test the claim of a supplier that their new osmium-tipped spark plugs can increase gas mileage by 5%. For each trial, the technician mounts a new set of either the standard spark plugs or the osmium-tipped spark plugs in a static mounted engine, and runs the engine for a certain simulated distance. The trials are run in random order, and yield the following data for gasoline consumed (measured in mL).

Plugs	Gasoline Consumption
Standard	548
Osmium	502
Osmium	530
Standard	552
Osmium	515
Osmium	515
Standard	553
Osmium	518
Standard	545
Osmium	527
Standard	535
Standard	553

(*a*) Plot the data in time order and look for trends. Would an upward or downward trend still affect this experiment since it was randomized?

(*b*) Analyze the experiment using analysis of variance. Is the difference between the two groups of spark plugs significant at the 0.05 level?

7. Some important properties of a metal film can be inferred from its reflective properties. A certain fab has two sputtering machines that are well matched for reflectivity of the films they deposit, but they are trying to qualify a new machine whose characteristics are not yet known. A process technician runs five wafers on each of the three machines, then measures the reflectivity on each wafer.

Machine 1	Machine 2	Machine 3
0.601	0.588	0.638
0.574	0.605	0.643
0.614	0.597	0.615
0.622	0.613	0.650
0.595	0.593	0.647

(*a*) Compare the means for the three machines. Are they significantly different?

(*b*) Compare the three means using a Tukey multiple comparison test.

★   (*c*) Suppose that the specification limits for this process ranged from 0.590 to 0.650, and the target was 0.620. The proportion of product which fails at the specification limit is 90%. Using an upside-down normal loss function (UDNLF) tailored to this situation and information obtained from the experiment, determine the expected loss in a stable metal deposition process that is centered on its target.

(*d*) What change in process centering would result in an expected loss increase of 5%?

(*e*) What sample size would be necessary to detect the difference determined in (*d*) if six etchers were to be involved in the experiment?

8. A competitive hiker climbs the same mountain each morning, always recording the time it takes him to reach the summit. In an effort to improve his times, he has traditionally eaten a very expensive carbohydrate bar, said to promote endurance, prior to climbing. On days when he is out of carbohydrate bars, he eats a banana, and notices that his times don't seem to change much. To test whether the carbohydrate bar is helping at all, he decides to alternate randomly between eating carbohydrate bars (1), bananas (2), and mesquite cheese curls (3) prior to his next 30 climbs. He records the following data in (Food, Time) order. Times are in minutes:

$$(3, 20.1), (3, 20.7), (1, 20.5), (2, 20.6), (1, 20.4),$$
$$(2, 20.2), (3, 20.4), (1, 20.1), (3, 20.4), (1, 20.7),$$
$$(1, 20.1), (2, 19.9), (3, 20.5), (2, 20.4), (1, 20.6),$$
$$(3, 20.7), (3, 21.2), (2, 20.7), (1, 20.5), (1, 20.3),$$
$$(2, 20.4), (3, 20.4), (2, 20.4), (2, 20.8), (3, 20.6),$$
$$(2, 20.8), (1, 20.3), (3, 20.6), (2, 20.5), (1, 21.1)$$

(*a*) Plot the data in time order. Does there appear to be any trend?

(*b*) Is there a significant difference between the treatments at a 0.05 significance level?

(*c*) How long would he have to continue the experiment in order to detect a 0.1-minute difference with alpha = 0.05, beta = 0.10?

9. A statistician living in the desert decides to plant a hedge of standard oleanders, a hardy plant for hot regions. He has a drip watering system which applies a precise amount of water to each plant on a periodic interval. Since he does not know the correct amount of water to give these plants, and since there are nine plants, he decides to vary the amount of water each will receive. Through a very easy modification to the drip system, he is able to set it so that three of the oleanders receive 1 gallon of water per hour, three receive 2 gallons per hour, and the final three receive 4 gallons per hour. At the beginning of the experiment, he records the height of each plant, and at the end of 90 days he again measures the height and thereby calculates how much each plant grew during the experiment. The growth, in centimeters, for each plant is given below:

Group	Growth
1 gallon	9
1 gallon	7
1 gallon	9
2 gallon	10
2 gallon	9
2 gallon	11
4 gallon	15
4 gallon	15
4 gallon	17

(a) Plot growth versus water rate for the 9 plants.
(b) Analyze the data using ANOVA. Are the groups significantly different?
(c) Perform linear regression to relate growth rate to water rate. Can you conclude that the slope is greater than zero? If so, find the slope and intercept.

10. An Olympic bobsled team desires to compare sled runners from three different manufacturers. They make 12 runs down the hill one morning, changing the runners each time in a randomized experiment. They record the following times, in seconds:

Manufacturer	Time
2	47.92
1	48.02
2	47.95
3	48.35
1	47.99
3	48.04
2	47.93
3	48.19
1	47.86
3	47.86
1	48.62
2	47.92

(*a*) Plot the residuals in time order. Does there appear to be any pattern?

(*b*) Compare the brands of runners using ANOVA.

(*c*) What would have been the likely result of this analysis if the manufacturer had not been randomized?

11. An engineer suspects that an ion implanter generates particles that cause implantation to be blocked in certain regions of the device, resulting in nonfunctional devices. There are three implanters in the equipment set, and he compares sort yield for 15 lots that went through the implanters in a 12-hour period. The figures are percent yields on a particular product.

Implanter 1	Implanter 2	Implanter 3
70	80	88
83	79	77
71	81	84
77	85	82
88	77	78

(*a*) Compare the means using ANOVA. If a significant difference exists, perform a Tukey multiple comparison test, and state the result.

(*b*) How many lots would the engineer need to run to detect a 2% difference in yield, with alpha = 0.05, beta = 0.10?

12. A physicist is studying high-temperature ceramic superconductors, and notices that adding small amounts of barium seems to increase the critical temperature at which superconductivity occurs. He makes up four batches of each of four recipes, which contain 1%, 1.5%, 2% and 2.5% barium, respectively. He measures the following critical temperatures, in degrees Kelvin, for the samples:

1%	1.5%	2.0%	2.5%
69	76	76	70
70	74	75	72
71	79	76	70
72	77	74	70

(*a*) Do the different concentrations of barium give different critical temperatures with a significance level of 0.05?

(*b*) Perform linear regression modeling critical temperature as a linear function of barium concentration.

(*c*) Plot critical temperature versus barium concentration. Is there a better model than the one described in (*b*)?

13. A slumping baseball player wants to test whether using a corked baseball bat (a bat in which the end has been hollowed out and filled with a foreign substance) will increase the distance he can hit a baseball. He will test two corked bats—one filled with cork, one filled with an elastic rubber substance—against a standard bat. To minimize variation he will hit balls pitched from a machine set to the slowest possible speed. He decides to hit 15 balls with each bat. To save time changing bats, he takes the 15 hits with the corked bat first, followed by 15 hits with the bat filled with rubber, followed by 15 hits with the standard bat. The distances of each hit, in feet, are given below.

corked bat: 364, 355, 380, 354, 379, 364, 364, 355, 360, 361, 365, 349, 373, 368, 336

rubberized bat: 382, 380, 363, 349, 378, 351, 367, 349, 358, 354, 371, 353, 358, 380, 342

standard bat: 341, 360, 353, 353, 353, 343, 335, 328, 338, 335, 334, 337, 362, 346, 340

(*a*) Do either of the altered bats give better performance than the standard bat at a significance level of 0.05?

(*b*) Plot the residuals in time order. Do you accept the conclusion of the ANOVA you performed?

(*c*) What could the baseball player have done to drastically improve the validity of his results?

14. A plumbing contractor wishes to test the strength of bonds in PVC piping made with cements from three different manufacturers. He begins by preparing six samples using each cement. Each sample consists of a 3-foot section of pipe with an end cap cemented at the end. Once the cement has dried, the samples are randomly tested on a pressure tester, which introduces water into the pipe and gradually raises the pressure until a failure occurs. The burst strengths, in pounds per square inch, are given below:

Vendor A	Vendor B	Vendor C
131	109	126
122	110	136
118	102	129
126	101	135
121	115	138
139	135	123

(*a*) Are the burst strengths different for the different vendors using a Tukey multiple comparison test?

(*b*) What sample size would be required to find a 4 psi difference in a follow up experiment with alpha risk of 0.50 and beta risk of 0.10?

15. A semiconductor fabricator uses silicon wafers from three different suppliers. Incoming wafers are inspected for a defect known a stacking faults. Since these defects follow a Poisson distribution, the defect densities can be compared using a chi-squared test, as in Section 8.6.1. A technician observes the following total defect counts on an equal area sample from each vendor:

   Vendor A: 26 defects
   Vendor B: 16 defects
   Vendor C: 19 defects

   Test the hypothesis that the three vendors have equal means for stacking faults.

16. Vendor B and Vendor C are being considered as sources for 300-mm silicon wafers, so there is interest in comparing just these two. Some additional data is now available for Vendor C also—an area equal to that originally inspected was found to have 11 defects. Use the method in Section 2.6.1 for comparing two defect densities to determine if these vendors are equivalent.

17. BarrV is one of the electrical tests performed after wafer processing which is known to have a nonnormal distribution, but the distribution is symmetric about its median so the Kruskal-Wallis test can be applied. Use this test to determine if the data obtained from three different plasma etchers provides convincing evidence that they produce different values for BarrV.

1	*2*	*3*
30.1	31.1	31.9
30.3	31.4	32.1
30.1	31.5	32.3
30.5	31.5	32.3
30.7	31.8	32.4
30.9	31.8	32.5

CHAPTER 3

BLOCKING

3.1 INTRODUCTION

This chapter introduces blocking, an experimental technique just as important as replication or randomization. Blocking is a boon to efficiency, allowing the experimenter to perform experiments with a fraction of the material that would otherwise be required. Failure to take advantage of blocking is probably the most common flaw of experiments in the semiconductor industry.

The format of this chapter mirrors that of Chapter 2: an introduction to blocking is provided in Section 3.2, design of blocked experiments is explained in Section 3.3, execution in Section 3.4, and analysis in Section 3.5. Section 3.6 covers an additional topic, restrictions on randomization.

3.2 BLOCKING

Blocking is a technique of experimentation that filters out unavoidable response variance so that the response can be more closely measured, much like a pair of sunglasses filters glare to give a clearer view of the scenery.

The principle behind blocking is readily demonstrated when analyzing data which are naturally paired. Suppose that a shoe manufacturer (Brand A) wished to prove that their new athletic shoes were more durable than those of a competitor (Brand B). They selected 10 basketball players and gave each of them one shoe of their own brand, and one of the competing brand. After 3 weeks of use, the shoes were recovered and tread wear measured. A boxplot of the differences in tread wear

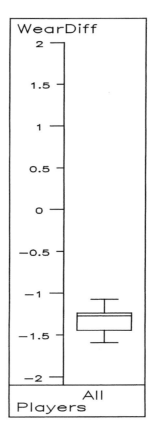

Fig. 3.1   This boxplot of the differences in tread wear for each player (A minus B) clearly shows that Brand A wears less than Brand B.

for each player (A minus B) is shown in Figure 3.1; obviously, Brand A shoes wear less than Brand B because the difference is negative.

Notice that the data chosen for analysis was the *difference* in shoe wear for each person—people have different weights, body builds, and running habits, so examining the wear difference for each player seems to make much more sense than treating the 20 tread wear measurements as independent observations. If this latter analysis method were chosen, the variances caused by player differences could easily hide the average tread wear difference experienced by each subject. Consider the boxplot of the 20 tread wear measurements in Figure 3.2, in which the wear difference is completely obscured.

Another (wrong) way to conduct this experiment would have been to give one group of players Brand A shoes, and the other group Brand B. If the first group of players happened to be all bigger and more active than the second group, then Brand A would have appeared to wear faster than Brand B—this is exactly the opposite result than should have been obtained.

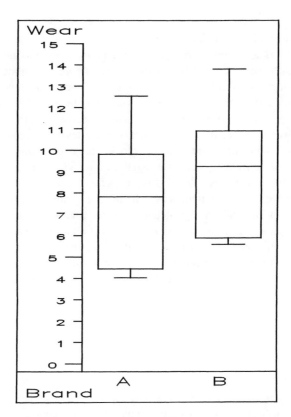

Fig. 3.2   The boxplot of tread wear for brand A is on the left, that for brand B is on the right. Even though there is a difference in tread wear between the brands, it cannot be seen because of the variance between individual players.

In this experiment, player was a blocking factor. A blocking factor is a source of response variance that adds or subtracts some offset to every experimental unit in the block, thus causing experimental units within the block to be more similar to one another than they are to experimental units in other blocks. The real factor of interest was the brand of shoe; this is called the *treatment factor.*

Blocking factors have several properties that distinguish them from most other sources of variance: they are usually known to the experimenter, they are often unavoidable, and treatments can be assigned to members of each block individually.

In nearly any fab process, lot is a block: wafers are almost always processed together as a lot, but lots are processed separately from other lots in many processing steps. Consequently, wafers within a lot tend to be more similar to one another than to wafers in other lots.

A variable may act as a blocking factor in some situations but not in others: people are a logical blocking factor in studies of shoe wear, but not (as diffusion tube operators) in comparisons of gate oxide thickness. In the first situation, people influence the response (shoe wear); in the second, human influence on automated diffusion equipment is very unlikely.

Blocking factors add variance to experiment responses, thus making it more difficult to find differences caused by factors of true interest (treatment factors). Blocked experiments explicitly recognize the effect of blocking factors, and reduce the variance they add by basing estimates of treatment factor effects only on differences observed within the same block. Blocking works because the effects of the blocks themselves are never allowed to influence estimates of the treatment effects. Instead, several small experiments (one per block) are done, treatment effect estimates are obtained from each experiment, and these effects are averaged to get overall treatment effects estimates. Blocking has the effect of making it easier to discern real differences due to treatments by reducing unexplained response variance.

Failure to block can obscure significant treatment effects, or make truly unimportant treatments appear to have a significant effect on a response. Blocked experiments avoid this type of destructive confounding, they make it easier to find true treatment factor effects, and they do all of this without any additional cost.

Two statistical models are presented in this text to describe blocked experiments. One model is:

$$y_{ijk} = \mu + \tau_i + \beta_j + \varepsilon_{ijk}$$

where:

y_{ijk} is the response for the kth replicate at the ith level of the treatment factor in the jth block. y must be a continuous; not categorical.

μ is the overall population mean.

τ_i is the effect of the ith level of the treatment factor. This is the amount of response added (or subtracted, if τ_i is negative) whenever the factor is at level i. Because these τ_i are offsets from an overall mean, their sum is constrained to be zero.

β_j is the effect of the jth block. (This β has no relation to beta risk; this is just an unfortunate notational coincidence.) Blocks are assumed to be a random effect, so β_j is a random sample from a normal distribution with mean zero and standard deviation σ_β.

ε_{ijk} is a random error for the kth replicate at the ith level of the factor in the jth block. These errors are assumed to be independent and normally distributed with mean zero and constant variance σ^2.

$i = 1 \ldots p$,   where p is the number of levels of the factor.

$j = 1 \ldots q$,   where q is the number of blocks.

$k = 1 \ldots n$,   where n is the number of replicates in each block at each level of each factor. If n is greater than 1, the model is said to be replicated; if n is equal to 1, the subscript is dropped and the model is said to be unreplicated.

Depletion Implant

Ion implantation is one means used to adjust the electrical properties of semi-conductors to desired levels. In concept an implanter is very simple: it is a machine which flings a particular type of ion into wafers in a controlled manner.

Ion implantation came to be used in the 1970s when diffusion was no longer sufficient to achieve doping goals. Diffusion required inorganic barriers like SiO_2 as masks, and used heat in excess of $400°C$. Ion implants can take place at room temperature utilizing organic (photoresist) or inorganic barriers as masks. The resultant doping is more precise both in terms of dose (number of atoms implanted per cm^3) and location. Because of the considerably lower temperature of implants, sideways travel of implanted atoms is much less likely than with diffusion doping.

The actual implant process consists of four steps that are illustrated in Figure 3.3:

1.   Produce some ions to implant. This is usually accomplished by exposing a gas or vaporized solid to a stream of electrons.
2.   Separate the desired ion from the others present in the ion chamber. This is done with a *mass analyzing magnet* or *separator,* which can very precisely select ions with mass equal to that of the desired ion.
3.   Move the ions fast enough to penetrate into the wafer, rather than bouncing off. An electrical potential moves ions from the separator toward the wafer—voltages of 100 keV (100,000 V) are not uncommon.
4.   Move the ion beam, which is only about 1 cm in diameter, over the surface of the wafer to evenly dope the entire wafer. The beam itself is *scanned,* just as an electron beam is moved across the screen of an oscilloscope. In many implanters, the wafers are simultaneously moved about to achieve better uniformity and faster throughput.

Wafers are implanted individually or in batches, depending on the particular equipment used. Dose can be very well controlled, and acceleration voltage can be controlled within 1 keV. Aside from initial setup, which includes beam focus, very little human interaction is required in the process.

In the experiments mentioned in this chapter, the threshold (turn-off) voltage of depletion transistors is being adjusted by doping the transistor channel with phosphorus.

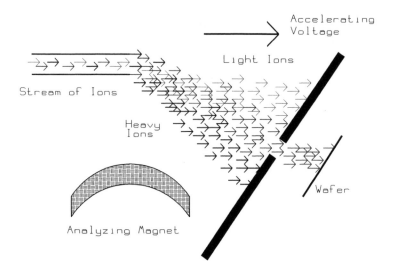

Fig. 3.3   This schematic view of the ion implant process shows the introduction of ions at the top left. These are accelerated toward a slit where ions of the chosen mass pass through and penetrate the wafer being implanted. Ions are sorted by the analyzing magnet (bottom left) which attracts heavier ions more than lighter ones.

■ *Example 3.1: An Implant Experiment Model*

Four implanters are to be involved in an experiment to see if the particular implanter employed during depletion implant makes a difference in the voltage threshold (V_T) as measured on a test structure. Three lots will be involved in the experiment, and two wafers from each lot will be run on each implanter. Aside from the assignment of wafers to lots (which is unavoidable), the experiment will be entirely randomized.

In this experiment, the implanters are a treatment factor with four levels, and the lots are a blocking factor. Two replicates are being run.

A statistical model representing this situation is:

$$y_{ijk} = \mu + \tau_i + \beta_j + \varepsilon_{ijk}$$

where:

y_{ijk} is the VT for the kth wafer run on the ith implanter in the jth lot.
μ is the overall population V_T.
τ_i is the effect of the ith implanter. This is the amount of response added or subtracted whenever that implanter is used.
β_j is the effect of the jth lot.
e_{ijk} is a random error for the kth wafer run on the ith implanter in the jth lot.
$i = 1 \ldots 4, j = 1 \ldots 3, k = 1 \ldots 2.$

One hidden assumption in the model above is that the effect of the treatment factor is the same in every block—the change in implant dose in the example above is expected to cause the same amount of threshold voltage change in every lot. Sometimes this extra assumption is not justified, and an interaction term must be included in the model:

$$y_{ijk} = \mu + \tau_i + \beta_j + (\tau\beta)_{ij} + \varepsilon_{ijk}$$

where $(\tau\beta)_{ij}$ is the interaction effect of the *i*th level of the factor with the *j*th block. The interaction effects are normally distributed with mean zero and variance:

$$\frac{p-1}{p}\sigma_{\tau\beta}^2$$

Interaction effects are assumed to be subject to the following constraint:

$$\sum_{i=1}^{p}(\tau\beta)_{ij} = 0 \quad for \quad j = 1 \dots q$$

These last two conditions were chosen to simplify the analysis of interaction models; there are alternate formulations that work equally well.

■ *Example 3.2: An Implant Experiment Model with Interaction*

An interaction term was added to the implant experiment model to account for the possibility that implanters interact with lots:

$$y_{ijk} = \mu + \tau_i + \beta_j + (\tau\beta)_{ij} + \varepsilon_{ijk}$$

where $(\tau\beta)_{ij}$ is the interaction effect of the *i*th implanter with the *j*th lot. There are a total of 12 interaction terms, 1 for each possible of combination of implanter and lot.

Model choice is the responsibility of the experimenter, and should be based on knowledge of the particular physical circumstances of each experiment. Interaction terms can often be left out of a model because interactions between blocks and factors are known to be very unlikely or so small that they will not substantially affect the analysis of the experiment.

3.3 DESIGN

The most critical step in designing blocked experiments is to recognize blocking factors. In almost any batch processing environment, the batches are a natural blocking factor. Some less obvious blocking factors are the several separate time periods in which a large experiment is performed, or processing chemical batches (resist, for example).

It seldom hurts to block: if the blocking factor has a significant effect on the response, a superior estimate of the treatment factor effect is obtained. If the blocking effect is not significant, it can be dropped from the analysis without incurring any additional cost. On the other hand, if the experimenter does not account for blocks, there is no way to reanalyze the experiment as if it had been blocked. Blocking does complicate an experiment a bit, but this small cost is repaid many times over by increased accuracy.

Experimental goals for blocked experiments are very similar to those in a simple comparative experiment: a statistical assessment of the effect of the treatment factor on a set of responses is desired. Experiments are designed to find a real difference of predetermined size (δ) with low risk of failure (β), while falsely indicating the existence of differences with low risk (α). The sample size necessary for such experiments can be found from tables similar to those used in the last chapter for simple comparative experiments.

Tests and estimates relating to block effects are of less interest—blocks are already assumed to have a significant effect on some response, or they would not have been included in the experiment.

Balance is even more important in blocked experiments than in simple comparative experiments. A blocked experiment is considered balanced if the same number of experimental units are assigned to every block-treatment factor combination. Balanced blocked experiments are more efficient than unbalanced experiments (more likely to find real differences), they are less affected by deviations from model assumptions, and the statistical tests needed for seriously unbalanced experiments are more intricate than those used for balanced experiments.

Block size is not always predetermined. If a time period (day, or shift) is chosen as a blocking factor, then the number of experimental runs conducted within each block can be controlled by the experimenter. An assumption is made in this text that at least one complete set of the treatment factor combinations can be run inside each block. This type of experiment is a *randomized complete block design*. Incomplete block designs are more difficult to execute and analyze; see Montgomery (1984) or Neter et al. (1985) for details.

Recall that alpha risk (α) is the probability of concluding that the treatment factor causes a difference when it does not actually affect the response, and beta risk (β) is the probability of failing to detect an actual difference in the response caused by the factor. In the sample size tables given here, alpha risk is fixed at 5% and beta risk is fixed at 10% with regard to a specific alternative (real difference) chosen by the experimenter. Experiments are assumed to be complete and balanced.

Table 3.1 gives sample sizes for a number of differences of interest, denoted by Δ at the top of each part of the table. The number of levels of the treatment factor, *p,* is shown at the top of each pair of columns in the body of the table. The number of blocks, *q,* is shown in the leftmost column of the table. A pair of sample sizes is shown under each column: the number on the left assumes that no interaction term is to be included in the model; the number on the right assumes that interaction terms will be included.

TABLE 3.1.   Sample Size for Blocked Analysis of Variance

$\Delta = 0.25$

					p								
	2		3		4		5		6		7		8
q													
2	169	1712	203	1441	228	961	247	819	264	759	280	730	293	716
3	113	493	136	325	152	302	165	298	176	301	187	306	196	312
4	85	202	102	177	114	179	124	184	132	190	140	196	147	202
5	68	124	82	122	91	127	99	133	106	139	112	144	118	149
6	57	90	68	93	76	99	83	105	88	110	94	114	98	119
7	49	70	58	76	65	81	71	86	76	91	80	95	84	99
8	43	58	51	64	57	69	62	73	66	78	70	81	74	85
9	38	49	46	55	51	60	55	64	59	68	63	71	66	74
10	34	43	41	49	46	53	50	57	53	60	56	63	59	66
11	31	38	37	43	42	48	45	51	48	54	51	57	54	59
12	29	34	34	39	38	43	42	46	44	49	47	52	49	54
13	26	31	32	36	35	39	38	42	41	45	43	47	46	49
14	25	29	29	33	33	36	36	39	38	41	40	44	42	46
15	23	26	28	31	31	34	33	36	36	38	38	41	40	42
16	22	25	26	29	29	31	31	34	33	36	35	38	37	40
17	20	23	24	27	27	29	30	32	32	34	33	35	35	37
18	19	22	23	25	26	28	28	30	30	32	32	33	33	35
19	18	20	22	24	24	26	26	28	28	30	30	32	31	33
20	17	19	21	22	23	25	25	27	27	28	28	30	30	31
21	17	18	20	21	22	24	24	25	26	27	27	28	28	30
22	16	17	19	20	21	22	23	24	24	26	26	27	27	28
23	15	17	18	19	20	21	22	23	23	25	25	26	26	27
24	15	16	17	19	19	20	21	22	22	24	24	25	25	26
25	14	15	17	18	19	20	20	21	22	23	23	24	24	25
26	13	15	16	17	18	19	19	20	21	22	22	23	23	24
27	13	14	16	16	17	18	19	20	20	21	21	22	22	23
28	13	13	15	16	17	17	18	19	19	20	20	21	21	22
29	12	13	14	15	16	17	18	18	19	19	20	20	21	21
30	12	12	14	15	16	16	17	18	18	19	19	20	20	21

$\Delta = 0.75$

					p								
	2		3		4		5		6		7		8
q													
2	20	191	23	161	26	107	28	91	30	85	32	82	33	80
3	13	55	16	37	18	34	19	34	20	34	21	34	22	35
4	10	23	12	20	13	20	14	21	15	22	16	22	17	23
5	8	14	10	14	11	15	12	15	12	16	13	16	14	17
6	7	10	8	11	9	11	10	12	10	13	11	13	11	14
7	6	8	7	9	8	9	8	10	9	11	9	11	10	11
8	5	7	6	8	7	8	7	9	8	9	8	9	9	10
9	5	6	6	7	6	7	7	8	7	8	7	8	8	9
10	4	5	5	6	6	6	6	7	6	7	7	7	7	8

TABLE 3.1.   Sample Size for Blocked Analysis of Variance (Continued)

11	4	5	5	5	5	6	6	6	6	6	6	7	6	7
12	4	4	4	5	5	5	5	6	5	6	6	6	6	6
13	3	4	4	4	4	5	5	5	5	5	5	6	6	6
14	3	4	4	4	4	4	4	5	5	5	5	5	5	6
15	3	3	4	4	4	4	4	4	4	5	5	5	5	5
16	3	3	3	4	4	4	4	4	4	4	4	5	5	5
17	3	3	3	3	4	4	4	4	4	4	4	4	4	5
18	3	3	3	3	3	4	4	4	4	4	4	4	4	4
19	3	3	3	3	3	3	3	4	4	4	4	4	4	4
20	2	3	3	3	3	3	3	3	3	4	4	4	4	4
21	2	2	3	3	3	3	3	3	3	3	3	4	4	4
22	2	2	3	3	3	3	3	3	3	3	3	3	3	4
23	2	2	3	3	3	3	3	3	3	3	3	3	3	3
24	2	2	2	3	3	3	3	3	3	3	3	3	3	3
25	2	2	2	2	3	3	3	3	3	3	3	3	3	3
26	2	2	2	2	3	3	3	3	3	3	3	3	3	3
27	2	2	2	2	2	2	3	3	3	3	3	3	3	3
28	2	2	2	2	2	2	2	3	3	3	3	3	3	3
29	2	2	2	2	2	2	2	2	3	3	3	3	3	3
30	2	2	2	2	2	2	2	2	2	3	3	3	3	3

$$\Delta = 1.00$$

q	p=2		p=3		p=4		p=5		p=6		p=7		p=8	
2	12	107	14	91	15	61	16	52	17	48	18	46	19	45
3	8	31	9	21	10	19	11	19	12	19	12	20	13	20
4	6	13	7	12	8	12	8	12	9	12	9	13	10	13
5	5	8	6	8	6	8	7	9	7	9	8	9	8	10
6	4	6	5	6	5	7	6	7	6	7	6	8	7	8
7	4	5	4	5	5	6	5	6	5	6	6	6	6	7
8	3	4	4	4	4	5	4	5	5	5	5	6	5	6
9	3	4	3	4	4	4	4	4	4	5	4	5	5	5
10	3	3	3	4	3	4	4	4	4	4	4	4	4	5
11	3	3	3	3	3	3	3	4	4	4	4	4	4	4
12	2	3	3	3	3	3	3	3	3	4	3	4	4	4
13	2	2	3	3	3	3	3	3	3	3	3	3	3	4
14	2	2	2	3	3	3	3	3	3	3	3	3	3	3
15	2	2	2	2	2	3	3	3	3	3	3	3	3	3
16	2	2	2	2	2	2	2	3	3	3	3	3	3	3
17	2	2	2	2	2	2	2	2	2	3	3	3	3	3
18	2	2	2	2	2	2	2	2	2	2	2	3	3	3
19	2	2	2	2	2	2	2	2	2	2	2	2	2	3
20	2	2	2	2	2	2	2	2	2	2	2	2	2	2
24	1	1	2	2	2	2	2	2	2	2	2	2	2	2
27	1	1	1	1	2	2	2	2	2	2	2	2	2	2
30	1	1	1	1	1	1	2	2	2	2	2	2	2	2

TABLE 3.1. Sample Size for Blocked Analysis of Variance (Continued)

$\Delta = 1.50$

	p=2		p=3		p=4		p=5		p=6		p=7		p=8	
q														
2	6	48	7	41	7	27	8	23	8	22	9	21	9	20
3	4	14	5	10	5	9	5	9	6	9	6	9	6	9
4	3	6	4	5	4	5	4	6	4	6	5	6	5	6
5	3	4	3	4	3	4	3	4	4	4	4	4	4	5
6	2	3	3	3	3	3	3	3	3	4	3	4	3	4
7	2	2	2	3	2	3	3	3	3	3	3	3	3	3
8	2	2	2	2	2	2	2	3	2	3	3	3	3	3
9	2	2	2	2	2	2	2	2	2	2	2	2	2	3
10	2	2	2	2	2	2	2	2	2	2	2	2	2	2
12	1	1	2	2	2	2	2	2	2	2	2	2	2	2
13	1	1	1	1	2	2	2	2	2	2	2	2	2	2
14	1	1	1	1	1	1	2	2	2	2	2	2	2	2
15	1	1	1	1	1	1	1	1	2	2	2	2	2	2
16	1	1	1	1	1	1	1	1	1	1	2	2	2	2
17	1	1	1	1	1	1	1	1	1	1	1	1	2	2
18	1	1	1	1	1	1	1	1	1	1	1	1	1	1

$\Delta = 2.00$

	p=2		p=3		p=4		p=5		p=6		p=7		p=8	
q														
2	4	27	4	23	5	16	5	13	5	12	5	12	6	12
3	3	8	3	6	3	5	3	5	4	5	4	5	4	5
4	2	4	2	3	3	3	3	3	3	3	3	4	3	4
5	2	2	2	2	2	2	2	3	2	3	2	3	3	3
6	2	2	2	2	2	2	2	2	2	2	2	2	2	2
8	1	1	1	1	2	2	2	2	2	2	2	2	2	2
9	1	1	1	1	1	1	1	1	2	2	2	2	2	2
10	1	1	1	1	1	1	1	1	1	1	1	1	2	2
11	1	1	1	1	1	1	1	1	1	1	1	1	1	1

Table 3.1 gives the sample size sufficient to detect a standardized difference in treatment effects of size Δ with 10% beta risk, assuming that alpha risk is fixed at 5%. For models without interaction, Δ is defined in terms of a difference δ (in the true scale of the data) and the within-group, within-block standard deviation, σ, as follows:

$$\Delta = \frac{\delta}{\sigma}$$

Δ has a different interpretation for models with interaction terms. The divisor in that case must include the variance component for block by treatment interaction:

$$\hat{\sigma} = \sqrt{\sigma^2 + n\sigma^2_{\text{Block*Treatment}}}$$

Table 3.1 gives sample sizes for a number of differences of interest, denoted by Δ at the top of each part of the table. The number of levels of the treatment factor, p, is shown at the top of each pair of columns in the body of the table. The number of blocks, q, is shown in the leftmost column of the table. A pair of sample sizes is shown under each column: the number on the left assumes that no interaction term is to be included in the model; the number on the right assumes that an interaction will be included.

Redundant lines have been omitted from the tables; use the values for the next lower q if the desired q values are not shown. For example, in the $\Delta = 1.00$ table, values for $q = 21, 22$, and 23 are not shown; use the values of $q = 20$.

Both the number of blocks and the size of each block must be chosen in a blocked experiment, so there are many possible experiments that could satisfy the same risk goals. Suppose an experiment with three treatment levels ($p = 3$) is desired which limits beta risk at $\Delta = 1.50$, and that no interaction term was to be included. According to the $\Delta = 1.50$ section of Table 3.1, this experiment could be conducted in two blocks ($q = 2$), each comprised of seven replications of all three treatment combinations. Each block would contain 21 observations, and the entire experiment would require 42 observations. The experiment layout is shown schematically below:

Level	1	2	3
Block			
1	7	7	7
2	7	7	7

If, for some reason, the experiment had to be run in five blocks ($q = 5$), then three observations per treatment combination per block would be needed, for a total of 45 observations:

Level	1	2	3
Block			
1	3	3	3
2	3	3	3
3	3	3	3
4	3	3	3
5	3	3	3

Some experiments do not fit exactly into blocks. For example, an experiment with seven treatment levels will never fit into a block of size 24 without some observations left over. There are several reasonable choices for the extra observations in the lot:

- Use the extra observations as centerpoints. Centerpoints are observations run at the usual process settings. The centerpoints can be deliberately spaced throughout the experiment as a means of monitoring uncontrolled variables.
- Use the extra observations to ensure that risky treatment levels are actually realized. Some levels may never have been tried before, so placing extra observations there helps ensure that at least some data will be obtained. This does result in an unbalanced experiment, but unless the unbalance is dramatic, it will not substantially affect experiment risks.
- Decrease the lot size so there is no waste. This is the most cost-effective alternative, but it does add risk to the experiment by making the experimental material different than usual process material. Some process steps may produce unusual results with small lot sizes, thus jeopardizing the entire experiment.

For models without interaction, Δ is defined as it was for simple comparative experiments:

$$\Delta = \frac{\delta}{\sigma}$$

where δ is the range of treatment effects at which beta risk is to be limited, and σ is an estimate of within-group within-block error. An estimate of σ can be obtained as the root mean square of within-block variances from normal production data.

If routine production data is unavailable, use the square root of mean square error (MSE) from an earlier experiment, as long as that experiment was blocked and replicates within blocks were obtained. Estimates from unblocked experiments will be much larger than necessary and will result in unnecessarily expensive experiments.

■ *Example 3.3: Finding Sample Size for an Experiment Without Interaction*

A depletion implant experiment like the one in Example 3.1 must be designed to detect a threshold Voltage (V_T) shift of 0.125 V. The within-lot standard deviation was found for two implanters that are routinely used for this implant: one standard deviation was 0.0751, and the other was 0.0943. These two estimates were combined as follows to arrive at an estimate for within-lot, within-implanter standard deviation:

$$\hat{\sigma} = \sqrt{\frac{(0.0751)^2 + (0.0943)^2}{2}} = 0.0852$$

The desired difference of 0.125 V is about 1.5 times this standard deviation estimate, so the $\Delta = 1.5$ portion of Table 3.1 will be used.

Four implanters are involved, so $p = 4$. The standard lot size for this factory is 12, so no more than three replicates may be obtained. The smallest number of blocks satisfying these constraints is found in the q row where the number of replicates is first three or less. Six blocks will suffice.

Blocking is used mainly because it is efficient: if blocks vary, then blocked experiments will require less material than unblocked experiments.

Consider an experiment which is to compare four machines in a setting with a fixed block size of 12. If a difference of size Δ is desired to be detected with only 10% beta risk, then the number of blocks required for the experiment are as shown below:

σ_B/σ	Blocks Needed If Blocking Is Recognized	Blocks Needed If Blocking is Not-Recognized
0	10	10
0.25	10	11
0.50	10	13
0.75	10	16
1.00	10	20
1.25	10	25

The first column of this table shows the ratio of the variance from blocks to the intrinsic process variance; the greater the ratio, the greater the advantage of blocking.

Δ has a different interpretation for models with interaction terms. The divisor in that case must include an estimate of block by treatment interaction:

$$\hat{\sigma} = \sigma^2 + n\sigma^2_{T*B}$$

Note that the number of replicates is included in this expression; this somewhat complicates the determination of sample size.

◼ *Example 3.4: Finding Sample Size for an Interactive Model*

Recall that in the previous example a difference of 0.125 V was chosen as the specific alternative to be detected 90% of the time. Experimenters learned from an old report that the lot by implanter interaction component of variance will probably be around 0.0442.

The lot size will require that three replicates be run (four implanters and lots of 12 wafers), so $n = 3$. The standard deviation to be used in computing Δ is now found as follows:

$$\hat{\sigma} = \sqrt{(0.0852)^2 + 3(0.0442)^2} = 0.115$$

and

$$\Delta = \frac{0.125}{0.115} = 1.087$$

so the $\Delta = 1.00$ portion of Table 3.1 will be used. Ten lots will be needed for the experiment.

Some experiment options do not allow for replication of treatment levels within blocks. When this is occurs, the interaction effect cannot actually be estimated, so the model is equivalent to one without an interaction term. This may seem to be a very efficient way to conduct an experiment, but it does have some risks. If interaction is actually present, its statistical significance cannot be tested, and it may give misleading results in the test for the treatment effect.

Replication within blocks also reduces the risk of losing an entire block-treatment level combination: without replication, it only takes one dropped wafer to lose an entire cell of the experiment.

3.4   EXECUTION

The random assignment of experimental units to treatment levels within each block is crucial: confounding variables can obscure or distort treatment effects in a blocked experiment just as they can in a simple comparative experiment.

■ *Example 3.5: Randomization in the Depletion Implant Experiment*

Another factory has four implanters, and uses eight-wafer lots. Three lots were split across the four implanters, and the order of the tests was randomized in the following manner. The numbers from 1 to 24 were written on slips of paper, mixed up in a hat, pulled out one at a time, and assigned to the cells of the experiment grid as follows:

	Imp-1		Imp-2		Imp-3		Imp-4	
Lot	Wafer	Run	Wafer	Run	Wafer	Run	Wafer	Run
1	5	7	1	10	6	18	2	4
	7	2	3	5	8	15	4	16
2	1	17	6	21	4	9	2	19
	7	13	8	22	5	11	3	12
3	5	8	3	6	2	1	1	24
	6	3	8	14	4	23	7	20

So wafer 5 in lot 2 was the 11th wafer implanted, and was implanted on machine 3.

Blocked experiments are a little more complicated than simple comparative experiments, and some of the designs introduced in following chapters are much more complex. As experiments become more elaborate, human error in execution can become as great a risk as alpha or beta risk. Some elements of successful experiments are:

- *Simplicity:* include only those factors and treatments which are necessary.
- *Uniformity:* if every block is treated the same way, they are all more likely to be treated correctly.
- *Ease:* make it easy to do the experiment right by requiring no more work than is absolutely necessary. Do not demand extra measurements or special handling unless they are needed to satisfy the goals of the experiment.
- *Clear instructions:* write absolutely clear instructions and make sure the right people have them when it comes time to process experimental material.
- *Redundancy:* Never count on one production lot or one wafer for essential data.

3.5   ANALYSIS AND INTERPRETATION

Analysis of blocked experiments is similar to that for simple comparative experiments: estimates of the treatment effect are obtained (Section 3.5.1), ANOVA is used to test for treatment factor significance (Section 3.5.2), and assumptions must be checked (Section 3.5.3). The calculations involved are more complex, and they vary depending on the model chosen, but modern statistical software makes most of the added complexity invisible to the experimenter.

Some manual computation methods are given for the ANOVA hypothesis tests in the figures which follow. These methods are valid for balanced completely randomized blocked experiments, but do not apply in any other situation.

3.5.1   Effects Estimates

Treatment factor effects are obtained exactly as they were before: the difference of the average response at each treatment level (from all blocks) from the overall average response estimates the effect at that level.

■ *Example 3.6: Treatment Effects Estimates*

Raw data from the implanter experiment of Example 3.5 is shown below, and represented graphically in Figure 3.4:

	Imp-1		Imp-2		Imp-3		Imp-4	
Lot	Wafer	V_T	Wafer	V_T	Wafer	V_T	Wafer	V_T
1	5	2.448	1	2.465	6	2.506	2	2.341
	7	2.359	3	2.516	8	2.336	4	2.447
2	1	2.362	6	2.468	4	2.285	2	2.224
	7	2.425	8	2.499	5	2.372	3	2.123
3	5	2.165	3	2.224	2	2.052	1	2.071
	6	2.248	8	2.151	4	2.035	7	2.069

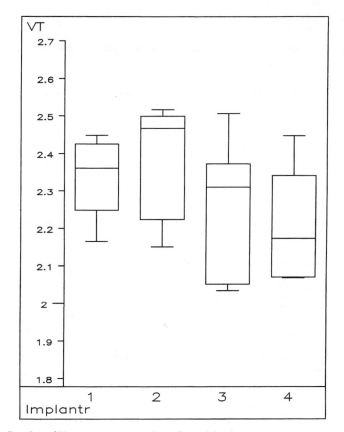

Fig. 3.4 Boxplots of V_T measurements are shown for each implanter.

So the wafer 5 in lot 2 was implanted on machine 3, and had a V_T of 2.372.

Implanter effects estimates were computed by finding the average V_T for the implanter, and then subtracting the overall response mean of 2.300:

Implanter	Average V_T	Estimated Effect
1	2.335	0.03488
2	2.387	0.08754
3	2.264	−0.03529
4	2.213	−0.08713

Interval estimates can be obtained by the Tukey method as before, with some minor corrections for interaction when it is assumed to be present.

■ *Example 3.7: Tukey Means Comparisons*

Tukey's procedure was run for the data from Example 9.5, with the edited results from SAS shown below:

```
         Tukey's Studentized Range (HSD) Test
                 for Variable: Vᴛ*

   Tukey Grouping        Mean        n        Imp.

                A        2.3872       6          2
                A
       B        A        2.3345       6          1
       B
       B        C        2.2643       6          3
                C
                C        2.2125       6          4
```

·Alpha = 0.05; df = 18; MSE = 0.00505; critical value of studentized range = 3.997; Minimum significant difference = 0.116. Means with the same letter are not significantly different.

The procedure identifies three groups of implanters: group A contains implanters 1 and 2, and is significantly different from group C, which contains implanters 3 and 4. Group B contains implanters 1 and 3, and this group is not significantly different from either of the other groups. This apparent contradiction is a common result of a means comparison procedure; it does not mean that any part of the analysis is incorrect.

Individual block effects can be estimated in the same way that treatment effects are, but they are inherently less interesting. Random blocks cannot be repeated, so the effect of any particular block cannot be repeated; it is only an indication of the variance that blocks add to the response.

■ *Example 3.8: Estimating Individual Block Effects*

For the data shown in Example 3.7, block effects were estimated in exactly the same manner as implanter effects, with the following results:

Lot	Mean V_T	Estimated Effect
1	2.427	0.1276
2	2.345	0.0451
3	2.127	–0.1728

These estimated effects reflect the differences shown in Figure 3.5.

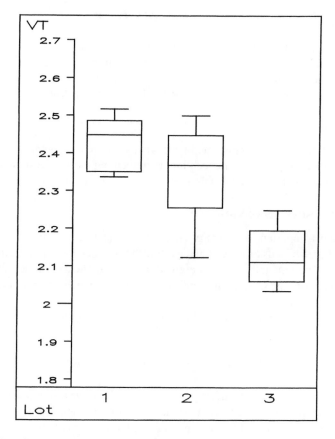

Fig. 3.5   The effect of lot is clear in these boxplots of V_T by lot.

Interaction effects are differences in treatment factor effects from one block to the next. They may be estimated as follows:

$$(\tau\beta)_{ij} \approx \bar{y}_{ij\cdot} - \bar{y}_{i\cdot\cdot} - \bar{y}_{\cdot j\cdot} + \bar{y}_{\cdots}$$

where $\approx$ indicates estimation.

The dot notation is shorthand for sums over dotted indices. For example,

$$\bar{y}_{ij\cdot} = \frac{1}{pn} \sum_{k=1}^{n} y_{ijk}$$

and

$$\bar{y}_{\cdot j\cdot} = \frac{1}{pn} \sum_{i=1}^{p} \sum_{k=1}^{n} y_{ijk}$$

■ *Example 3.9: Estimating Individual Interaction Effects*

The interaction of the second lot with the third implanter will be estimated from the data of Example 3.5. The average V_T for that cell of the table is 2.329, the average for the third implanter is 2.264, the average for the second lot is 2.345, and the overall average is 2.300. The interaction is estimated as follows:

$$(\tau\beta)_{32} \approx 2.329 - 2.264 - 2.345 + 2.300 = 0.0200$$

Interaction effects are most interesting when they are large—this may indicate some anomaly in experiment execution, or something worth investigating about the nature of block-to-block differences.

3.5.2   Analysis of Variance

The most important outcome of ANOVA for blocked experiments is the hypothesis test that leads to a decision regarding the significance of the treatment factor effect. Three models will be presented here: the replicated model without an interaction term, the model with replication and an interaction term; and the unreplicated model.

In each of these models, the main hypotheses of interest are:

$$H_0: \tau_1 = \tau_2 = \ldots = \tau_p = 0$$
$$H_A: \tau_i \neq 0 \quad \text{for some} \quad i \in \{1, 2, \ldots, p\}$$

If H_0 is rejected, then the treatment factor is said to have a significant effect on the response. If H_0 is not rejected, then the effect was not statistically significant—it could be zero, or it could simply be too small to detect with that particular experiment.

3.5.2.1   *The replicated model without interaction.*   Blocked experiments are easy to analyze with statistical software, as is seen in the SAS example that follows.

■ *Example 3.10: ANOVA with SAS*

Edited SAS output for the data of Example 3.5 is shown below:

General Linear Models Procedure—Dependent Variable: V_T

Source	df	Sum of Squares	F Value	Pr > F
Lot	2	0.38533575	38.15	0.0001
Imp.	3	0.10629646	7.02	0.0025
Error	18	0.09089742		
Corrected total	23	0.58252963		

"Pr > F" is the probability that the observed F-ratio would occur if the null hypothesis were true, so a small value (less than 0.05) should lead the experimenter to reject the null hypothesis.

The F-tests indicate that both Lot and Imp (implanter) have a significant effect on threshold voltage (VT). The F-test for Lot is best used for indication only—wafers were not randomly assigned to lots, so there may be some question about the validity of the test.

If the blocking factor had not been recognized, implanters would not have appeared to have a significant effect on voltage thresholds. This is demonstrated with the analysis below of the same data, but without the blocking factor:

General Linear Models Procedure—Dependent Variable: V_T

Source	df	Sum of Squares	F Value	Pr > F
IMP	3	0.10629646	1.49	0.2481
Error	20	0.47623317		
Corrected total	23	0.58252963		

The analysis of blocked experiments is like standard ANOVA except that the sum of squares due to blocks (SSB) must be estimated. The complete ANOVA table is shown in Table 3.2. The first column names the source of variance, the second gives a formula for finding the sum of squares associated with that source. The dot notation is used throughout these formulas to signify summarization over dotted subscripts. For example,

TABLE 3.2.   ANOVA Table for Replicated Blocked Model Without Interaction*

Source	SS	df	MS	F
Treatment	$\dfrac{1}{qn}\sum_{i=1}^{p} y_{i\cdot\cdot}^2 - \dfrac{y_{\cdot\cdot\cdot}^2}{pqn}$	$p-1$	MST	$\dfrac{\text{MST}}{\text{MSE}}$
Blocks	$\dfrac{1}{pn}\sum_{j=1}^{q} y_{\cdot j\cdot}^2 - \dfrac{y_{\cdot\cdot\cdot}^2}{pqn}$	$q-1$	MSB	$\dfrac{\text{MSB}}{\text{MSE}}$
Error	$SSCT - SST - SSB$	$pqn-p-q+1$	MSE	
Corrected total	$\sum_{i=1}^{p}\sum_{j=1}^{q}\sum_{k=1}^{n} y_{ijk}^2 - \dfrac{y_{\cdot\cdot\cdot}^2}{pqn}$	$npq-1$		

*This complete analysis of variance (ANOVA) table for blocked experiments without interaction terms shows the sources of variance (first column), the sum of squares (SS), and degrees of freedom associated with that source (second and third columns). SSCT is the sum of squares for the corrected total. The mean square (MS) is the SS divided by the degrees of freedom. The *F*-test appropriate for each source is in the last column of the table.

$$y_{\cdot j} = \sum_{i=1}^{p} y_{ij}$$

The next column of the table gives the degrees of freedom associated with the source; this is used to compute the mean square due to blocks (MSB) from the sum of squares due to blocks as follows:

$$\text{MSB} = \frac{\text{SSB}}{q-1}$$

The last column is the most important, because it describes the *F*-test that will be used to make a decision about the significance of effects. The ratio to be tested is listed; the degrees of freedom for the *F*-percentile are those associated with the numerator and denominator of the *F*-ratio. For example, the treatment *F*-ratio will be compared with the $F_{p-1,\,pqn-p-q+1,\,0.95}$ percentile.

The hypothesis test has only two outcomes: it either indicates that the factor has a (statistically) significant influence on the response, or it does not. Either of these outcomes can occur erroneously if the ANOVA assumptions are violated, so the actions outlined in Section 3.5.3 are critical to ensuring the validity of conclusions.

If the factor effect is significant, level-specific effects should be estimated and means comparison methods used to see which levels differ from the overall mean, and how. In some cases, the magnitude of the factor effect may not be of practical

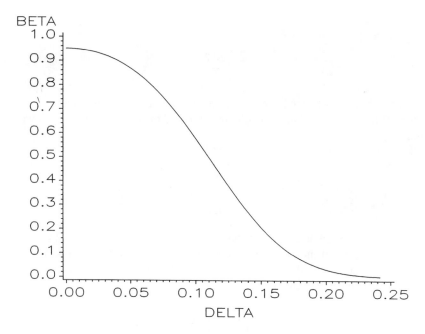

Fig. 3.6   This beta risk curve shows the risk of failing to detect differences (DELTA) in experiments similar to the one already conducted.

significance. Although statistically significant, it may be so small that no reasonable person would act to change the process to eliminate the effect.

If the factor effect does not appear to be significant and the model assumptions have been verified, then a beta risk curve (Figure 3.6) can be used to quantify the probability that a factor effect did exist, but was not detected. Although this post hoc examination of beta risk should be unnecessary if the experiment was designed to find differences of practical importance, the beta risk curve is still very useful when explaining experimental risks and results to others. Beta risk curves are essential in the evaluation of experiments where design goals were not documented.

The block effect is usually assumed to be significant, but this assumption can be refuted by an *F*-test. If a block effect does not appear to be significantly different from zero, then there may be an opportunity to simplify future experiments by ignoring blocks. If the blocks were small and many blocks were involved in the experiment, a substantial increase in the power of the test for a significant treatment effect will also result.

For example, if 10 blocks of size three were used in an experiment involving a treatment effect with three levels, 18 degrees of freedom for error are available for the test of the treatment effect. The ratio used to test the treatment effect is:

$$\frac{\left(\dfrac{SST}{2}\right)}{\left(\dfrac{SSE}{18}\right)}$$

If blocks were found to be insignificant, and were left out of the analysis, 27 degrees of freedom would be available for error, and the ratio used in the test would be:

$$\frac{\left(\dfrac{SST}{2}\right)}{\left(\dfrac{SSE}{27}\right)}$$

The sum of squares due to random error (SSE) is not exactly the same as the previous one, but if the block effect is insignificant, the two will be very similar and the second ratio will be larger than the first. However, the ratios are compared to very similar critical values:

$$F_{2,\,18,\,0.95} = 3.59$$
$$F_{2,\,27,\,0.95} = 3.35$$

So in this case, leaving blocks out of the analysis furnishes a much more sensitive test of the treatment effect.

3.5.2.2   *The replicated model with interaction.* Analysis of interactive models differs from that required for noninteractive models because the *F*-ratio used to test treatment and block effects is based on an estimate of interaction, not on random error. As seen in Table 3.3, the correct ratio for testing treatment effects is:

$$F_0 = \frac{\dfrac{SST}{p-1}}{\dfrac{SST*B}{(p-1)(q-1)}}$$

where SST is the sum of squares due to the treatment and SST*B is the sum of squares due to treatment by block interaction.

This differs from the usual (noninteractive model) test, which uses:

$$F_0 = \frac{\dfrac{SST}{p-1}}{\dfrac{SSE}{pqn-p-q+1}}$$

TABLE 3.3   ANOVA Table for Replicated Blocked Model with Interaction*

Source	SS	df	MS	F
Treatment	$\dfrac{1}{qn}\sum\limits_{i=1}^{p} y_{i\cdot\cdot}^2 - \dfrac{y_{\cdots}^2}{pqn}$	$p-1$	MST	$\dfrac{MST}{MST*B}$
Blocks	$\dfrac{q}{pn}\sum\limits_{j=1}^{q} y_{\cdot j\cdot}^2 - \dfrac{y_{\cdots}^2}{pqn}$	$q-1$	MSB	$\dfrac{MSB}{MSE}$
Interaction	$\dfrac{1}{n}\sum\limits_{i=1}^{p}\sum\limits_{j=1}^{q} y_{ij\cdot}^2 - \dfrac{y_{\cdots}}{} - SST - SSB$	$(p-1)(q-1)$	MST*B	$\dfrac{MST*B}{MSE}$
Error	SSCT – SST – SSB – SST * B	$pq(n-1)$	MSE	
Corrected total	$\sum\limits_{i=1}^{p}\sum\limits_{j=1}^{q}\sum\limits_{k=1}^{n} y_{ijk}^2 - \dfrac{y_{\cdots}^2}{pqn}$	$pqn-1$		

*This ANOVA table for replicated blocked experiments with interaction terms gives the correct ratio for testing the treatment factor effect.

where SSE is the sum of squares due to random error.

Using the wrong test will give the wrong answer, and to make matters worse, some software will choose the wrong test by default; the user must specify the correct ratio to ensure the correct test is used.

The correct analysis is shown in the example below.

■ *Example 3.11: SAS Analysis of an Interactive Model*

Another set of data was collected in an experiment similar to that of Example 3.5: three eight-wafer lots were split across four implanters. The measurements obtained are shown below:

	Implanter			
Lot	1	2	3	4
1	2.448	2.465	2.506	2.341
	2.359	2.516	2.336	2.447
2	2.582	2.608	2.285	2.224
	2.645	2.639	2.372	2.123
3	2.165	2.224	1.902	1.881
	2.248	2.151	1.885	1.879

An interaction plot (Figure 3.7) gives a strong indication that there is significant block by treatment interaction; this is proven by the statistical test below:

General Linear Models Procedure—Dependent Variable: V_T

Source	DF	Sum of Squares	F Value	Pr > F
IMP	3	0.35773313	5.08	0.0438
LOT*IMP	6	0.14097625	6.57	0.0029
Error	12	0.04293450		
Corrected total	23	1.34943296		

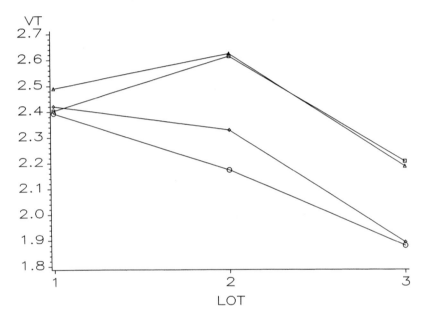

Fig. 3.7   This interaction plot indicates that lot-implanter interaction may be significant. The implanters seem to fall into two groups with respect to this interaction, as indicated by the two pairs of similar lines.

Note that the LOT*IMP interaction was used as the denominator for the *F*-test of implanter significance, and the Error term was used to test the LOT*IMP interaction itself.

Large block effects are not uncommon; in fact, they are expected. Large interaction effects are more rarely encountered and when they do occur, further investigation is warranted.

Interaction plots can often give valuable insight to the nature of block by treatment interactions. In one common pattern, one or two blocks show interactions much larger than the others (Figure 3.8). This probably means that there is something unusual about the odd blocks: they were processed incorrectly, the treatments were not applied as they should have been in those blocks, or they were different from others before being processed.

Another common pattern is for blocks to be grouped somehow in the nature of their interaction effects (Figure 3.9). This indicates the presence of a *lurking variable;* this is a variable that influences the response, but was not accounted for in the experiment. Starting material (silicon vendor) is a likely suspect in many ex-

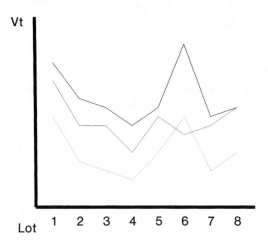

Fig. 3.8   In this interaction plot, only one of the blocks (lot number 6) appears to be unusual: it may have been processed incorrectly, the treatments might not applied as they should have been in that block, or it was different from the others before being processed.

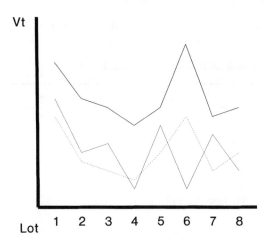

Fig. 3.9   In this interaction plot, even- and odd-numbered blocks seem to fall into two classes with respect to the effect of the treatment. Some lurking variable is probably causing this consistent pattern.

periments, but other factors like queue times between processing steps may also be guilty. If records are available to identify this lurking variable, then a reanalysis with the additional variable in the model can give a much clearer picture of true factor effects.

3.4.2.3 The unreplicated model.   The unreplicated model is economical, and it provides a valid test of the treatment effect. It provides no information on block-treatment interactions.

The model is given by

$$y_{ij} = \mu + \tau_i + \beta_j + \epsilon_{ij}$$

where:

$\quad y_{ij}$ is the response the ith level of the treatment factor in the jth block. y must be continuous, not categorical.

$\quad \mu$ is the overall population mean.

$\quad \tau_i$ is the effect of the ith level of the treatment factor. This is the amount of response added (or subtracted, if τ_i is negative) whenever the factor is at level i. Because these τ_i are offsets from an overall mean, their sum is constrained to be zero.

$\quad \beta_j$ is the effect of the jth block. (This β has no relation to that used to denote beta risk.) Blocks are assumed to be a random effect, so β_j is a random sample from a normal distribution with mean zero and standard deviation σ_β.

$\quad \epsilon_{ij}$ is a random error at the ith level of the factor in the jth block. These errors are assumed to be independent and normally distributed with mean zero and constant variance σ_2.

$i = 1 \ldots p,$ where p is the number of levels of the factor.

$j = 1 \ldots q,$ where q is the number of blocks.

The analysis of variance table for this model is shown in Table 3.4. A sample analysis is shown in Example 3.12.

TABLE 3.4.   ANOVA Table for Unreplicated Blocked Model*

Source	SS	df	MS	F
Treatment	$\dfrac{1}{q} \sum\limits_{i=1}^{p} y_{i\cdot}^2 - \dfrac{y_{\cdot\cdot}^2}{pq}$	$p-1$	MST	$\dfrac{\text{MST}}{\text{MSE}}$
Blocks	$\dfrac{1}{p} \sum\limits_{j=1}^{q} y_{\cdot j}^2 - \dfrac{y_{\cdot\cdot}^2}{pq}$	$q-1$	MSB	$\dfrac{\text{MSB}}{\text{MSE}}$
Error	SSCT − SST − SSB	$(p-1)(q-1)$	MSE	
Corrected total	$\sum\limits_{i=1}^{p} \sum\limits_{k=1}^{q} y_{ij}^2 - \dfrac{y_{\cdot\cdot}^2}{pq}$	$pq-1$		

*The complete ANOVA table for the unreplicated blocked model is shown here. Interaction of blocks with treatments cannot be tested here, but the test for the treatment effect is still valid even if interaction is present.

■ *Example 3.12: Analysis of Unreplicated Blocked Experiment*

Three four-wafer lots were split across four implanters, with the results shown below:

		Implanter		
Lot	1	2	3	4
1	2.448	2.516	2.336	2.341
2	2.284	2.390	2.431	2.372
3	2.113	2.250	2.123	1.975

The *F*-test found no significant difference due to implanters:

```
           General Linear Models Procedure—Dependent
                        Variable: V_T
```

Source	df	Sum of Squares	F Value	Pr > F
LOT	2	0.20429600	17.77	0.0030
IMP	3	0.03783158	2.19	0.1897
Error	6	0.03449667		
Corrected total	11	0.27662425		

The beta risk curve for this experiment (Figure 3.10) demonstrates the main disadvantage of unreplicated experiments: large differences may pass undetected. According to this curve, experiments of this type have only half a chance of detecting a 0.21 V difference in implanters.

3.5.3   Checking Assumptions

All the usual ANOVA assumptions need to checked for models with blocking, and as before, residual analysis is the most powerful tool available for this purpose. A *residual* in a blocked experiment is the difference between a the value predicted by the model and an actual observation. For the model with interaction, the residual is found as follows:

$$e_{ijk} = y_{ijk} - \bar{y}_{ij\cdot}$$

For the model without interaction terms, residuals are determined as follows:

$$e_{ijk} = y_{ijk} - \bar{y}_{i\cdot\cdot} - \bar{y}_{\cdot j\cdot} + \bar{y}_{i\cdot\cdot}$$

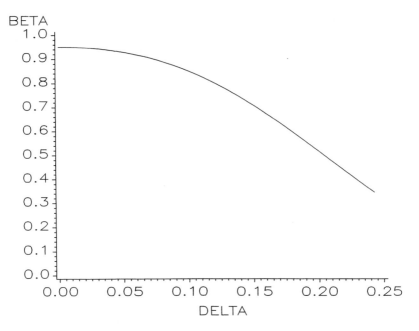

Fig. 3.10   The beta risk curve for this unreplicated blocked experiment demonstrates the main disadvantage of unreplicated experiments: large differences may pass undetected. According to this
curve, experiments of this type have only half a chance of detecting a 0.21-V difference in implanters.

■ *Example 3.13: Residual Computation*

Consider the data from Example 3.5. For the model with interaction, the residual
for the first observation (from the first wafer processed in lot 1 on implanter 1),
would be:

$$e_{111} = y_{111} - \bar{y}_{11.}$$
$$= 2.448 - 2.404 = 0.044$$

The first term is the actual observation, and the second is the mean of all measurements from lot 1 and implanter 1.

The residual computation for the model without interaction is a little different:

$$e_{111} = y_{111} - \bar{y}_{1..} - \bar{y}_{.1.} + \bar{y}_{...}$$
$$= 2.448 - 2.335 - 2.427 + 2.300 = -0.014$$

A histogram of residuals (Figure 3.11) should appear to be normally distributed. Observations corresponding to unusually large residuals should be investigated.

Boxplots of residuals by treatments (Figure 3.12) and by blocks (Figure 3.13) should appear to be generally uniform in spread.

A timeplot of residuals should still appear random even if the blocks were periods of time (Figure 3.14).

3.6 RESTRICTIONS ON RANDOMIZATION

An experiment is completely randomized if experimental units are randomly assigned to treatment combinations. The experiments in Chapter 2 were completely randomized experiments in one factor; in every case the experimental units (usually wafers) were randomly picked to be treated with levels of the factor. The number of wafers to be treated with each treatment level was not random—it was carefully controlled—but the assignment of wafers to treatments was random.

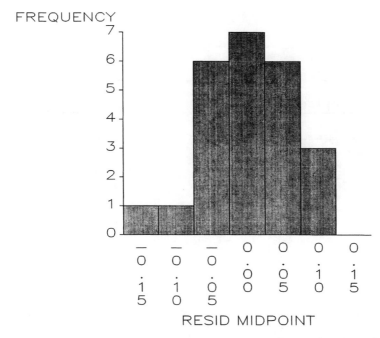

Fig. 3.11 This histogram of residuals from a blocked experiment seems to verify that the residuals are normally distributed.

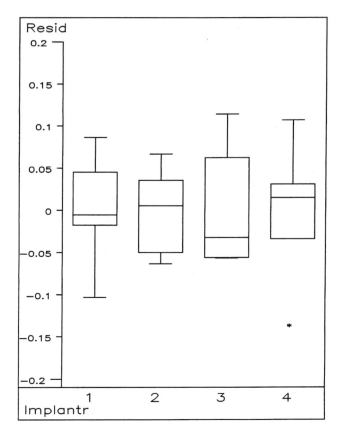

Fig. 3.12   Boxplots of residuals by implanter do not show significant heterogeneity of variance.

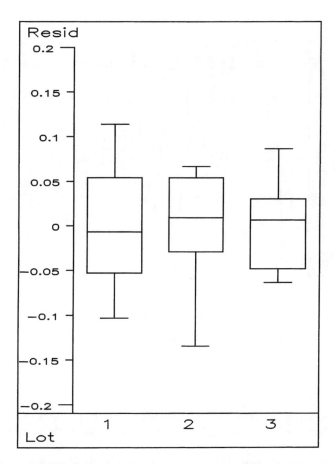

Fig. 3.13 Boxplots of residuals by lot seem to verify that variance with lots is uniform.

Blocking is one example of a restriction on randomization. A blocked experiment has two factors—one treatment factor and one blocking factor—and experimental units by their very nature reside in one and only one block. Hence, the random assignment of experimental units to treatment-block combinations is impossible.

Blocked experiment designs are a way to explicitly recognize a restriction on randomization, and produce valid experimental results in spite of the restriction. Two other types of experiment that also admit restrictions on randomization are explained below: the Latin square and the crossover design.

3.6.1 Latin Square Designs

Latin square designs are often applied when there are two natural blocking factors, and experiment resources are constrained so that only one replicate of each

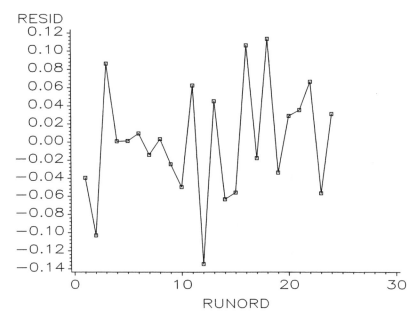

Fig. 3.14   This plot of residuals in time order does not reveal any obvious trends or other systematic patterns.

treatment level can be applied to each combination of the blocking factors. In this case, the number of blocks for each blocking factor is equal to the number of treatment factor levels.

Consider an experiment to evaluate five different types of cleanroom headgear. The headgear vendor will donate only one set of each type of headgear for the test, so only five people can be involved in the experiment at any one time. Each of the five subjects will wear one type of headgear for a week, and then fill out an evaluation form which assigns that headgear a score from 0 to 100 based on factors like visibility, comfort, and ease of use. The next week, each subject will try another type of headgear. The test will be complete at the end of 5 weeks, when every subject has tried every type of headgear exactly once.

The treatment factor in this experiment is the type of headgear. One blocking factor is subject; the other blocking factor is time. Note that there are five treatment levels, five subjects, and 5 weeks. The experiment is shown schematically below. Subjects form the columns of the tables, and weeks the rows. The five different types of headgear are denoted by the letters *A* through *E*.

			Subject		
Week	*1*	*2*	*3*	*4*	*5*
1	A	B	C	D	E
2	E	A	B	C	D
3	D	E	A	B	C
4	C	D	E	A	B
5	B	C	D	E	A

Note that each type of headgear appears only once in each row and column. The model for this Latin square is given by the following equation:

$$y_{ijk} = \mu + \tau_i + \beta_j + \gamma_k + \epsilon_{ijk}$$

where:

y_{ijk} is the response for the ith headgear as measured by the jth subsect in the kth week.

μ is the overall population mean.

τ_i is the effect of the ith headgear. Because these τ_i are offsets from an overall mean, their sum is constrained to be zero.

β_j is the effect of the jth subject. Subjects are assumed to be a random effect, so β_j is a random sample from a normal distribution with mean zero and standard deviation σ_β.

γ_k is the effect of the kth week. Weeks are assumed to be a random effect, so γ_k is a random sample from a normal d istribution with mean zero and standard deviation σ_γ.

ϵ_{ijk} is a random error for the ith headgear type as measured by the jth subject in the kth week. These errors are assumed to be independent and normally distributed with mean zero and constant variance σ^2.

$i = 1 \ldots, 5, \ j = 1 \ldots, 5, \ k = 1 \ldots, 5.$

The layout above is a standard latin square; the actual design was randomized in two steps before the experiment was performed. The first randomization step reordered rows by randomly selecting an order for the rows listed. Row 4 was moved to row 1, row 2 remained where it was, row 5 became row 3, row 3 became row 4, and row 1 was moved to row 5:

		Subject			
Week	*1*	*2*	*3*	*4*	*5*
1	C	D	E	A	B
2	E	A	B	C	D
3	B	C	D	E	A
4	D	E	A	B	C
5	A	B	C	D	E

The next randomization step reordered columns: column 2 become column 1, column 4 became column 2, column 1 became column 3, column 3 became column 4, and column 5 remained where it was. The result is the design as it was actually executed:

		Subject			
Week	*1*	*2*	*3*	*4*	*5*
1	D	A	C	E	B
2	A	C	E	B	D
3	C	D	B	D	A
4	E	B	D	A	C
5	B	D	A	C	E

Raw data from the experiment is shown next to the headgear type in the table below:

		Subject			
Week	*1*	*2*	*3*	*4*	*5*
1	D,60	A,67	C,63	E,77	B,36
2	A,67	C,28	E,65	B,45	D,44
3	C,35	D,48	B,54	D,69	A,54
4	E,78	B,49	D,62	A,75	C,19
5	B,55	D,57	A,81	C,51	E,40

The analysis of this design proceeds almost exactly like that of the design with one blocking factor. The important F-test is the one which tests the significance of the treatment factor. Edited SAS output including this test is shown below:

General Linear Models Procedure—Dependent Variable: SCORE				
Source	df	Sum of Squares	F Value	Pr > F
Week	4	365.36000000	1.72	0.2101
SUBJ	4	2412.16000000	11.36	0.0005
HEADGR	4	2734.56000000	12.87	0.0003
Error	12	637.28000000		
Corrected total	24	6149.36000000		

The F-tests shown in the last column of the printout clearly indicate that there are differences in headgear. Tests for week and subject are also provided, although they are of less interest because they were already presumed to be significant.

Computational details were omitted here; the interested reader can find more information on the analysis in Anderson and McLean (1974).

One disadvantage of the Latin square design is that interaction cannot be tested. Interaction plots like the one in Figure 3.15 can be used as a visual check for interaction. In this case, the lines are nearly parallel, so interaction between headgear and subjects is probably not present. Interaction between headgear and weeks, or between weeks and subjects, would have to be checked with corresponding interaction plots.

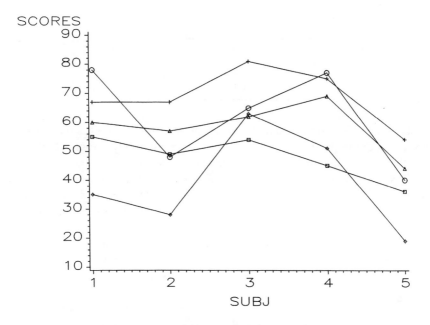

Fig. 3.15 The Latin square design does not allow for a statistical test of interaction, but this interaction plot does provide a visual check for interaction. Evaluators are plotted on the *x*-axis and headgear scores are on the *y*-axis. Each type of headgear is plotted on a separate line with a distinguishing symbol (circles for headgear 5, for example). The lines are nearly parallel in most cases, so interaction between evaluators and headgear types is not indicated.

3.6.2 Crossover Designs

The crossover design is useful when a treatment with two levels is applied to a number of subjects which act as blocks.

Suppose that an experiment is to be done to check for differences between two electrical test machines. The tests that are to be done can only be done on real product wafers, and those wafers vary considerably from one to another so they form a natural blocking factor.

The wafers must be tested on both testers, but the testers are in two different factories several hundred miles away so any single wafer must wait for some period of travel time between tests. Time is another blocking factor in this experiment, because some of the electrical properties of the wafers can spontaneously change over time. The process of wafer transportation can also change some electrical properties, especially if the wafers are placed on an aircraft and subjected to cosmic radiation. Since time and travel are perfectly confounded in this experiment—time only passes during travel—they act together as a single blocking factor.

The effect of time and travel could be great enough to hide differences due to the testers, so a crossover design was used: half the wafers in the experiment were tested on Tester 1 first, then on Tester 2; the other half of the wafers were tested on Tester 2 first, then on Tester 1. The experimenter was careful to have the wafers travel in the same manner and time of day between factories. The experiment layout is shown below:

	Day 1	Day 2
Tester 1	Wafers 1–5	Wafers 6–10
Tester 2	Wafers 6–10	Wafers 1–5

The crossover design essentially cancels out the time effect: even if there were a substantial time or travel effect, it would be spread evenly across both testers, and thus could not influence the estimate of the tester effect.

This crossover model depends on two assumptions. The first is that there is no residual effect from the tests—that testing on one tester does not influence later tests. The second assumption is that there is no interaction between the subjects, the treatments, or the time periods.

The model for this crossover design is given by the following equation:

$$y_{ijk} = \mu + \tau_i + \beta_j + \gamma_k + \varepsilon_{ijk}$$

where:

y_{ijk} is the response for the ith tester as on the jth wafer on the kth day.

μ is the overall population mean.

τ_i is the effect of the ith tester. Because these τ_i are offsets from an overall mean, their sum is constrained to be zero.

β_j is the effect of the jth wafer. Wafers are assumed to be a random effect, so β_j is a random sample from a normal distribution with mean zero and standard deviation σ_β.

γ_k is the effect of the kth time period. Time is assumed to be a random effect, so γ_k is a random sample from a normal distribution with mean zero and standard deviation σ_γ.

ε_{ijk} is a random error for the ith tester as measured on the jth wafer in the kth time period. These errors are assumed to be independent and normally distributed with mean zero and constant variance σ^2.

$i = 1 \ldots, 2, j = 1 \ldots, 5, k = 1 \ldots, 2$.

The analysis of this design proceeds in a manner very similar to that of the Latin square. The important F-test is the one that tests the significance of the tester factor. Edited SAS output including this test is shown below:

General Linear Models Procedure–Dependent Variable: YIELD

Source	DF	Sum of Squares	F Value	Pr > F
TESTER	1	6.05000000	0.36	0.5633
ORDER	1	61.25000000	3.68	0.0914
WAFER	9	7268.45000000	48.50	0.0001
Error	8	133.20000000		
Corrected total	19	7468.95000000		

The p-value for the testers is larger than 0.05, so the hypothesis that the testers are equal cannot be rejected on the basis of this data. Tests for the time periods (ORDER, here), and the wafers are also given.

3.7 SUMMARY

Recognizing blocking factors and using analysis methods to account for them is essential in nearly any batch processing environment.

Blocking saves resources by allowing differences to be found with much less experimental material than would be required with unblocked experiments.

Blocking prevents mistaken conclusions: unless blocking is used, important effects can be hidden, and unimportant effects can appear to be very significant.

CHAPTER 3 PROBLEMS

1. A professor of sociology suspects that his five teaching assistants (TAs) are giving different grades to students who are performing comparably. To test this hypothesis, he divides up 30 papers, giving 6 to each TA to grade. They come back with the following grades:

    TA 1: 99,93,91,86,94,89
    TA 2: 98,92,81,78,97,91

TA 3: 81,95,88,97,85,89
TA 4: 95,86,92,85,90,82
TA 5: 79,98,95,94,80,90

(*a*) Can he conclude that there is a difference among the TAs using ANOVA?
(*b*) Can you think of a better way to evaluate possible differences in grading standards?

2. A professor of physics suspects that his five teaching assistants are giving different grades to students who are performing comparably. To test this hypothesis, he makes photocopies of six tests and gives all six to each TA. They come back with the following grades:

TA	Test 1	Test 2	Test 3	Test 4	Test 5	Test 6
1	68	91	77	88	88	53
2	72	94	78	90	90	57
3	69	89	77	91	91	55
4	74	93	80	91	91	54
5	65	87	78	86	86	50

(*a*) Test the hypothesis that the TAs give equivalent grades using simple ANOVA.
(*b*) Test the same hypothesis by blocking on test number. How does this affect the result?

3. A lithography engineer is investigating the effect of exposure energy on critical dimensions (CDs) for a new type of resist. She knows that each wafer involved in the experiment will be coated and processed a little differently than the others, and so she has decided to use wafers as a blocking factors. Three patterns are printed at random locations on the wafer for each of four exposures—a total of 12 patterns are printed on every wafer. Three wafers were used for the experiment. The CD data she collected is shown below:

W	E	CD
1	90	2.40, 2.62, 2.47
1	95	2.55, 2.55, 2.19
1	100	2.11, 2.17, 2.36
1	105	2.06, 2.28, 2.34
2	90	2.37, 2.37, 2.60
2	95	2.13, 2.31, 2.35
2	100	2.33, 2.48, 2.29
2	105	2.46, 2.22, 2.07
3	90	2.26, 2.51, 2.32
3	95	2.4, 2.61, 2.22
3	100	2.23, 2.15, 2.35
3	105	2.48, 2.11, 1.99

(*a*) Compute point estimates for the exposure and block effects. Do they appear to be statistically significant?

(*b*) Construct an interaction plot (wafer by exposure) and note any apparent interactions.

(*c*) Do an ANOVA (with blocking) and test the significance of the exposure effect.

(*d*) Compute the residuals and plot them by wafer, then by exposure setting. Are there any obvious violations of the model assumptions?

(*e*) Using information from the data provided, determine the number of wafers which would be needed to detect an exposure effect of 0.10 microns (0.10 μm).

(*f*) If up to 30 patterns could be printed on each wafer, how many wafers would be needed for the experiment designed in (d).

(*g*) If up to 2000 patterns could be printed on each wafer, would it be reasonable to run the entire experiment on one wafer?

4. Before bottling, beer is "primed" by adding a small amount of sugar or malt extract. This is fermented by the yeast still living in the beer, and provides the carbonation which gives beer its head when poured; see Papazian (1991) for details.

An amateur brewer who is trying out three different brands of malt extract in the priming step knows that "batch" is a blocking factor: each 5-gallon batch of beer which is brewed at one time is different from every other batch. He brewed four batches, and primed one third of the bottles with each brand of malt extract. Bottles were randomly assigned from within each batch to an extract brand. Four bottles from each batch-brand combination were poured from special fixture designed for this purpose, and the height (H, in cm) and duration (D, in minutes) of the head was measured. Results are shown below:

BATCH	BRAND	H	D	H	D	H	D	H	D
1	1	4.7	2.20	5.4	2.45	6.8	2.40	3.0	0.95
1	2	3.8	1.65	3.6	1.40	5.8	2.00	3.5	1.45
1	3	3.8	1.95	3.7	1.25	2.6	1.25	3.2	1.40
2	1	4.4	1.60	5.5	2.70	3.8	1.15	3.5	1.45
2	2	2.7	0.90	5.2	1.85	3.6	1.60	4.7	2.00
2	3	4.1	1.50	5.9	2.10	2.0	0.35	4.6	1.90
3	1	3.4	1.40	6.2	2.80	6.2	2.30	4.5	1.80
3	2	3.4	1.85	2.3	1.25	6.1	1.95	4.6	2.05
3	3	3.7	1.75	2.4	1.05	3.9	1.75	1.8	0.45
4	1	4.8	1.75	4.6	1.65	5.3	2.10	5.7	2.15
4	2	5.1	2.60	5.8	2.55	2.6	1.05	5.6	2.30
4	3	6.5	2.45	4.1	1.70	3.6	1.40	6.4	2.95

(*a*) Compute point estimates for the brand and batch effects (for both responses). Do they appear to be statistically significant? Do the responses appear to be related?

(*b*) Do an ANOVA (with blocking) and test the significance of the brand effect for both responses.

(*c*) Compute the residuals for the head height and plot them by batch, then by brand. Are there any obvious violations of the model assumptions?

(d) Using information from the data provided, determine the size of experiment (batches and number of bottles which would have to be measured) that would be needed to detect a head height effect of 1 cm. You may assume that 4.5 gallons of the batch will actually be bottled from each batch, and that the beer will be put in 16-ounce bottles.

(e) Construct an interaction plot (batch by brand) for both responses and note any apparent interactions.

5. The Arizona Department of Transportation (ADOT) routinely uses blocking to measure the effectiveness of different types of paint. Stripes of different paints are painted perpendicular to traffic flow very near to one another at several locations around the state. Traffic patterns will be almost exactly the same within each location (a block) but will differ from one location to the next.

   Data on reflectivity (shown at right) was collected from 11 different types of paint at four different locations:

   (a) Compute point estimates for the location and paint effects. Do they appear to be statistically significant?

   (b) Make an interaction plot. Do you notice anything that would lead you to eliminate some of the data from this experiment?

   (c) Do an ANOVA (with blocking) on that portion of the data that you think will provide useful information, and test the significance of the paint effect.

6. You find yourself in a difficult position at your new job with HSC: your boss does not understand the concept of blocking, and thinks it is a waste of time to block by production lot. Prepare a 1-minute discourse to educate this person on the value of blocking.

7. A distance runner keeps track of the time it takes to run a fixed course, but he has noticed that his timings vary by season. He wants to try to assess the effect of three different brands of running shoe, so he looked at the following timings which were taken over a period of 6 months:

Month	Shoe	Timings			
1	1	22.90	24.25	23.30	24.10
1	2	24.15	22.00	21.80	22.20
1	3	22.35	20.85	22.20	22.50
2	1	21.45	21.45	22.85	20.30
2	2	21.40	21.70	21.90	22.80
2	3	20.40	21.75	20.35	19.45
3	1	19.55	21.05	19.90	20.70
3	2	21.95	19.65	20.00	19.55
3	3	19.30	20.45	18.25	17.50
4	1	18.85	19.75	19.35	20.10
4	2	19.25	18.85	20.75	18.80
4	3	18.20	18.50	18.90	18.45
5	1	18.85	18.10	18.60	17.50
5	2	19.70	17.15	17.30	16.55
5	3	17.10	16.45	14.75	16.35
6	1	15.60	16.55	17.35	17.65
6	2	17.35	15.80	16.30	16.40
6	3	14.75	13.30	14.50	16.15

Location	Paint	Reflectivity			
1	1	1.00	1.00	1.00	1.00
1	2	1.00	1.00	1.00	1.00
1	3	1.00	1.00	1.00	1.00
1	4	1.00	1.00	1.00	1.00
1	5	1.00	1.00	1.00	1.00
1	6	1.00	1.00	1.00	1.00
1	7	1.00	1.00	1.00	1.00
1	8	1.00	1.00	1.00	1.00
1	9	0.87	0.90	0.84	0.82
1	10	0.85	0.88	0.86	0.88
1	11	0.86	0.85	0.90	0.85
2	1	0.56	0.57	0.58	0.57
2	2	0.58	0.55	0.57	0.54
2	3	0.60	0.53	0.53	0.51
2	4	0.58	0.56	0.51	0.56
2	5	0.53	0.56	0.58	0.59
2	6	0.58	0.54	0.55	0.55
2	7	0.56	0.52	0.56	0.60
2	8	0.54	0.58	0.60	0.50
2	9	0.56	0.56	0.52	0.56
2	10	0.56	0.58	0.54	0.56
2	11	0.57	0.57	0.53	0.52
3	1	0.53	0.51	0.51	0.50
3	2	0.50	0.49	0.51	0.51
3	3	0.52	0.50	0.50	0.55
3	4	0.54	0.57	0.55	0.54
3	5	0.49	0.51	0.54	0.52
3	6	0.51	0.49	0.56	0.50
3	7	0.52	0.52	0.51	0.51
3	8	0.56	0.55	0.56	0.54
3	9	0.50	0.50	0.52	0.50
3	10	0.48	0.47	0.55	0.49
3	11	0.54	0.56	0.55	0.52
6	1	0.50	0.45	0.43	0.47
6	2	0.53	0.50	0.48	0.49
6	3	0.51	0.50	0.45	0.46
6	4	0.48	0.49	0.50	0.52
6	5	0.48	0.48	0.48	0.50
6	6	0.45	0.50	0.49	0.49
6	7	0.49	0.49	0.41	0.46
6	8	0.50	0.44	0.51	0.47
6	9	0.43	0.45	0.47	0.46
6	10	0.49	0.47	0.47	0.54
6	11	0.49	0.47	0.47	0.46

Perform a complete analysis of this data, treating month as a blocking factor.

★ 8.   Dogs have their puppies in litters, and the characteristics of the offspring vary more from one litter to the next than they do within each litter, hence litter is treated as a blocking factor in many experiments involving dogs. A dog obedience school is interested in testing four methods of training dogs to bark at intruders, but they have only four trainers who are qualified to train with these methods. Both litter and trainer are blocking factors here, and the training method is the single experimental factor of interest. A Latin square experiment will be conducted to test the effect of training method.

Dogs from four litters were trained and tested in a standardized intruder test—number of barks was the response. The results of this experiment are shown below in a Latin square layout: in each cell, training method is shown, followed by barks. Trainers are shown at the top of the table, so when Trainer 2 used method 2 to train a dog from litter 3, 308 barks were counted during the intruder test.

Litter	T1	T2	T3	T4
1	3,87	4,2	1,0	2,318
2	4,2	1,0	2,322	3,76
3	1,0	2,308	3,74	4,1
4	2,346	3,86	4,2	1,0

(a) Does the training method have an effect on intruder response, as measured by barking?
(b) What conclusions can you make about differences between litters?
(c) What conclusions can you make about differences between trainers?

9.   A new wafer sort floor is about to qualify their first tester for production. The standard qualification procedure requires that eight wafers be sorted on this tester, and on an equivalent machine in another factory. A crossover design must be used.

Yield data was collected on eight wafers randomly selected from the production line, with the results shown below. The first half of the wafers were sorted on tester 1 first, then on tester 2; the second half was sorted in the opposite order.

Tester	Sort	Yield			
1	1	217	246	228	243
2	2	226	254	236	250
2	1	233	253	228	257
1	2	235	253	230	259

(*a*) Is the new tester equivalent to the one in the other factory?

(*b*) Would you have arrived at this same conclusion if the crossover design had not been used?

10. An amateur brewer is interested in determining the effect of two brands of hops on the fragrance of a particular type of ale. Forty tasters will be involved in this experiment, and a crossover design will be used to cancel any carryover effects from one fragrance to another. Fragrance was rated on a scale from 1 to 10, with the results shown below:

Brand of Hops	*Tasting*	*Tasting Results Listed in Order by Taster—20 Tasters Sampled the Beer with Brand A First; 20 Sampled Brand B First*									
A	1	31	50	39	50	50	13	7	16	42	8
		37	43	34	35	50	8	31	38	43	50
B	2	29	50	40	50	50	11	7	15	40	8
		37	42	35	31	50	7	32	38	40	50
B	1	29	43	12	31	30	37	5	0	37	15
		18	8	13	7	22	22	29	16	13	48
A	2	50	50	35	50	50	50	28	21	50	41
		40	31	37	31	46	47	50	37	34	50

(*a*) Does the brand of hop have an effect on fragrance?

(*b*) Are all tasters similar in their perception of this effect?

(*c*) Suppose that three brands of hops needed to be evaluated; suggest a design that would be effective in finding their effects.

CHAPTER 4

FACTORIAL EXPERIMENTS

4.1 INTRODUCTION

Factorial experiments simultaneously estimate the effects of two or more experimental factors. They are usually done after screening experiments, so most factors included in the experiment are expected to have some effect on the response. Factorial experiments have impressive economic benefits because they allow the experimenter to measure the influences of many factors in a single experiment. Factorial experiments are especially valuable because they have the ability to detect factor *interactions*—those differences in the effect of one factor caused by the influence of other factors.

The design of a factorial experiments is simple: they are multifactor experiments in which every possible combination of factor levels is included in the experiment. Consider a tea brewing experiment in which there are two factors: water temperature and steeping time. Each factor has two levels. The layout of one replicate of the experiment is shown below. Temperature is in degrees Celsius; steeping time is in minutes.

Temperature	Steeping Time
45	4
45	360
90	4
90	360

Strength

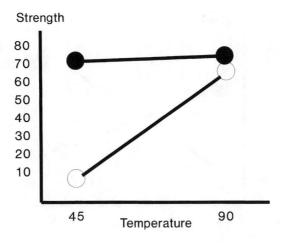

Fig. 4.1   In this interaction plot, the strength of tea (*y*-axis) depends on both the brewing, temperature (*x*-axis) and the brewing time. The results of two different brewing times are represented by separate lines. The effect of time is very different depending on the temperature—at the higher temperature, time does not have much effect on strength, but at the lower temperature, time has a very large
effect.

The response of interest in this experiment is the strength of the tea, as measured by color intensity. Results from one replicate of this experiment are shown
below:

Temperature	Steeping Time	Strength
45	4	8
45	360	72
90	4	68
90	360	76

As seen from the interaction plot in Figure 4.1, the effect of time is very different depending on the temperature—at the higher temperature, time does not
have much effect on strength, but at the lower temperature, time has a very large
effect.

The existence of interactions is the most compelling reason to use factorial experiments. If the effect of time had been evaluated in a separate experiment with
temperature fixed at 45°, a very large effect would have been observed (72 – 8 =
64), and the experimenter may have reached the erroneous conclusion that time always has an effect on strength. If time had been varied with temperature fixed at
90°, hardly any effect would have been seen (76 – 68 = 8). Neither experiment

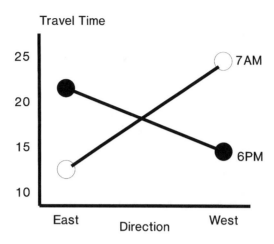

Fig. 4.2   This interaction plot clearly shows that both time of day and travel direction have a profound effect on travel time, but the effect of one of these factors depends on the setting of the other factor.

would give as full an explanation of the relationships between the factors and the response as the factorial.

Consider another factorial experiment with two factors: the first factor is the direction in which I drive between home and work, and it can take the values "East" and "West". The second factor is the time of day I make the drive, which can take the values "7 AM" and "6 PM". The response of interest is the time it takes to make the trip. One replication of the experiment is shown below with typical responses (in minutes).

Direction of Travel	Time of Day	Travel Time
East	7 AM	13
East	6 PM	22
West	7 AM	25
West	6 PM	15

The difference between the average travel time driving East (12.5 minutes) and the average travel time driving West (15 minutes) makes the effect of travel direction seem rather small. Similarly, the apparent effect of the time of day seems small: 19 minutes in the morning minus 18.5 minutes in the evening. The interaction plot in Figure 4.2 tells an entirely different story: both time of day and travel direction have a profound effect on travel time, but these effects would be hidden in experiments that deliberately varied only one factor at a time while allowing the other factor to vary freely.

A different type of deceptive result would be obtained if time of day were fixed and direction was varied—if the time of day happened to be morning, then West would appear to be slower than East; if evening were picked, then East would appear slower than West.

Interactions are very common in the semiconductor industry: many experimental factors interact, and the magnitude of their interaction can be nearly as large as the effect of any of the individual factors. Because factor interactions can only be detected if the factors are simultaneously varied, these important effects cannot be found by the simple comparative experiments described in Chapters 2 and 3.

The fundamental principles of design, execution, and analysis for factorial experiments are similar to those already presented: Sections 4.2, 4.3, and 4.4 demonstrate the basics of design, execution, and analysis for factorial experiments. Later sections develop these ideas and treat special situations like blocking and analysis of covariance (ANCOVA).

4.2 DESIGNING FACTORIALS

Factorial experiments are used to discover the effects of a number of factors, as well as any factor interactions, and to quantify those effects to some extent. Once the factors and their levels are selected, designing a factorial experiment is essentially the same as designing a simple comparative experiment: specify acceptable experimental risks, and then select an experiment that meets those goals.

A shorthand notation is used to describe factorial experiments that describes the number of factors, the number of levels for each factor, and the number of runs per replicate concisely: A k^p factorial experiment has p factors, each of which are varied to k levels, and k^p runs are required per replicate.

Thus a 2^4 experiment has four factors, each of which is varied to two levels. Sixteen runs are needed per replicate. If the entire experiment is replicated twice, 32 total runs will be required.

There are 3^p experiments (p factors each taken to three levels), and even more complicated factorials like the $2^3 \cdot 4^5$ experiment in which three factors are varied to two levels each, while simultaneously five factors are varied to four levels each. This last experiment would require 8,192 runs for a single replicate.

Factorial experiments with more than a few factors are very large: a 2^5 factorial has 32 runs per replicate, and a 2^8 factorial has 256 runs per replicate. Large factorials are difficult to execute correctly in a factory, and there are screening designs that can estimate main effects and simple interactions for a large number of factors much more efficiently than full factorials. For these reasons, full factorials with more than five factors are rarely seen in practice.

TABLE 4.1.   Replicates Required for a 2^k Factorial Experiment to Detect a Difference of Δ Standard Deviations with 10% Beta Risk*

					Δ					
k	0.25	0.50	0.75	1.00	1.25	1.50	1.75	2.00	2.25	2.50
2	169	43	20	12	8	6	5	4	3	3
3	85	22	10	6	4	3	3	2	2	2
4	43	11	5	3	2	2	2	2	2	2
5	22	6	3	2	2	2	2	2	2	2
6	11	3	2	2	2	2	2	2	2	2
7	6	2	2	2	2	2	2	2	2	2
8	3	2	2	2	2	2	2	2	2	2
9	2	2	2	2	2	2	2	2	2	2

*This table shows the number of replicates required for a 2^k factorial experiment to detect a difference of Δ standard deviations 90% of the time. The number of factors in the experiment is found in the k column; the differences are found under the Δ heading. If an experiment with three factors is required to detect a difference of 0.75 standard deviations, then 10 replications of the entire experiment (a total of 80 observations) will be required

For the estimation of linear effects, two levels per factor are sufficient, and for that reason, this chapter is focused on two-level experiments. Including more than two levels per factor results in very large experiments. A 2^4 factorial has 16 runs per replicate; a 3^4 factorial has 81. If more than two levels are truly needed—to estimate quadratic effects, for example—there are usually more efficient designs available.

Once the number of factors and levels for the factors are known, a number of replicates must be chosen. As with simple comparative experiments, the number of replicates depends on the experimental risks that are to be accepted. Use Table 4.1 to determine the number of replications *of the entire factorial* that will be likely to satisfy the experimental goals.

■ *Example 4.1: Choosing Sample Size for a Factorial Experiment*

A 2^4 factorial which is designed to detect (90% of the time) a difference of 1.25 standard deviations in the response will require two replications. Since each replicate of the 2^4 takes 16 runs, the entire experiment will require 32 runs.

A 2^5 factorial designed to detect a difference of 0.75 standard deviations requires three replications. Since the 2^5 experiment takes 32 runs per replicate, 96 runs will be needed for the entire experiment.

While some of these experiments may seem large, they are actually small when compared to those that would have to be used if each factor were investigated independently. For example, if the five factors in the 2^5 experiment above were investigated independently with the same experimental risks, then according to Table 2.1, 15 replicates of a simple comparative experiment would be required for each factor. Since each experiment in that setting would vary a factor to two levels, 30 runs would be needed for every one of these five experiments, for a total of 150 runs. This is more than a 50% increase in size compared to the 96 runs needed for the 2^5, *and* the separate (one factor at a time) experiments are incapable of detecting interactions.

Blocking factors are just as important in factorial experiments as they are in simple comparative experiments. They complicate matters a bit because large factorials sometimes will not fit into a natural block. There are ways to split factorials across blocks that are explained in Section 4.7.2, but this practice complicates the experiment and thus increases the risk that it will be executed incorrectly. A safer choice is to select factorial designs that will fit inside blocks.

Centerpoints are even more useful in factorials than in simpler experiments. They serve as an experimental control, as a source of within-group variance estimates in unreplicated experiments, and as a basis for testing model adequacy.

4.3 EXECUTING FACTORIALS

Factorials should be randomized like any other experiment; unfortunately, the complexity of these experiments encourages even well-intentioned experimenters to be a bit lazy. Consider the experience of Ted, Bob, and Enrico in their 2^2 experiment investigating the effects of a catapult machine setting (Rear Bandfix Height, or RBH) and operator of the machine (OPER) on the distance traveled by a cork after launching it from the catapult.

Rather than randomize the experiment, they ran an entire set of five replicates at one RBH-OPER combination, changed either OPER or RBH, and then ran five more replicates at these same settings, with results as shown (Figure 4.3). They no-

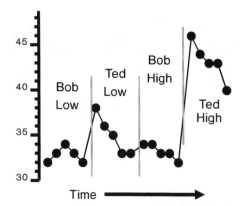

Fig. 4.3   Raw data from a catapult experiment is shown in this plot. Operator and RBH are noted for each part of the experiment: "Bob, Low" means that Bob was the operator, and that RBH was set to its low value. Interaction seems to be present because the "Ted, High" combination produces much higher values than would be expected to be caused by either factor.

ticed immediately that there appeared to be a very large interaction between operator and setting, which in this setting was quite incongruous—these factors should have been entirely independent.

The shift in the response came about when, during the course of the experiment, Enrico said to Ted, "Ted, you're letting that thing go too fast. You should hold it for a second before releasing it." Ted made the suggested change with obviously disastrous (for this experiment) results.

Recognizing this fact, Ted repeated the last set of five runs using his earlier release method, with the results shown in Figure 4.4. Unfortunately, it seemed that Ted was unable to recover his earlier technique to produce readings near those that were expected. Enrico suggested that Ted may have changed his stance since the earlier part of the experiment, and that it may have affected the results.

Ted modified his stance to what Enrico and Bob thought it was during the early part of the experiment, but still produced five runs with measurements much higher than expected. Bob then remarked that he had been measuring the distance up until Ted starting repeating parts of the experiment, and that Enrico had made all measurements since then.

Ted made five more runs with Bob measuring, but still produced implausible results. Enrico then suggested that Ted repeat the earlier part of the experiment as well to see if the results matched, but this suggestion was not taken to heart. They decided to repeat the entire experiment in randomized order at a later date.

This experiment took 35 runs, and no useful information was obtained.

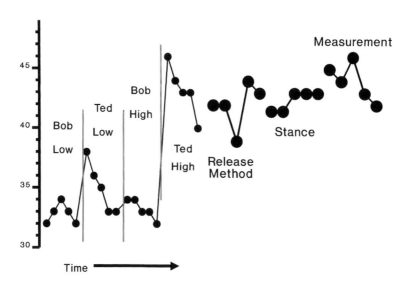

Fig. 4.4 Further trials with the catapult experiment could not produce values similar to the first part—changing release method, stance, and measurement method were all tried, but without success.

The Lithography Develop Process

Experiments in this chapter involve a part of the lithography process: the develop step. The develop step occurs near the end of the process as follows:

1. *Coat:* put resist on the wafer.
2. *Queue time:* some (carefully controlled) time lapses between coat and expose.
3. *Expose:* expose areas of the resist that will later be developed away.
4. *PEB:* some processes use a postexposure bake to stabilize the resist, or to make it less liable to changes caused by differing delay times from expose to develop.
5. *Queue time:* some (carefully controlled) time passes between expose and develop.
6. *Develop:* dissolve and rinse away those areas of resist that were exposed.
7. *Etch:* chemically remove underlying layers not protected by resist.
8. *Strip:* remove any remaining resist.
    The develop process itself can be complex. The process referenced in this chapter proceeds as follows:
6a. *Spin:* place the wafer horizontally on a vacuum chuck and spin it at a specific speed.
6b. *Spray:* developer is sprayed onto the wafer while it is spinning.
6c. *Develop:* some time is allowed for the develop to remain on the wafer.
6d. *Rinse:* all developer is rinsed off while the wafer spins.
6e. *Bake:* an optional bake may be required depending on which types of processing follow.

Together with exposure energy, develop parameters are the main determinants of line widths. Good control of line widths is essential for successful semiconductor processing: these critical dimensions (CDs) influence transistor thresholds, signal delay times, and influence device reliability.

Measuring the effects of the develop process can be difficult. Optical microscopes are often inadequate for this purpose because the wavelength of the light they use is too near the width of the lines being measured. Scanning electron microscope (SEM) measurements are highly accurate, but very expensive and time consuming.

One excellent surrogate for direct measurement is measuring the resistivity of long, thin lines etched from the pattern determined by the developed resist. A doped polysilicon serpentine (Figure 4.5) is used for this purpose. Together with a larger polysilicon area which is used to determine the bulk resistivity of the material, the serpentines are printed at hundreds of sites on a wafer. Critical dimension measurement is then simply a matter of measuring the resistance from start to end of each serpentine, and then using the bulk resistivity measurement to compute the width of the polysilicon line.

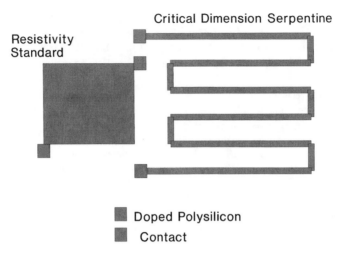

Fig. 4.5   A doped polysilicon serpentine provides an indirect measure of polysilicon CD. The measurement is based on the ratio of the serpentine resistance to the bulk resistivity.

4.4   ANALYSIS OF FACTORIALS

Factorials are analyzed in a manner similar to simple comparative experiments: factor effects are estimated (Section 4.4.1), factor significance is assessed (Section 4.4.3), and assumptions much be checked (Section 4.4.4). Some additional complications result from the fact that many factor interactions must be taken into account; interaction interpretation is the subject of Section 4.4.2.

The factorials considered here are two-level fixed-effect models. The model for a two-factor experiment would be:

$$y_{ijk} = \mu + \tau_i + \beta_j + (\tau\beta)_{ij} + \varepsilon_{ijk}$$

where y_{ijk} is the response at the kth replicate of the ith level of the first factor and the jth level of the second factor. The sum of effects for each factor is assumed to be zero:

$$\sum_{i=1}^{2} \tau_i = 0$$

$$\sum_{j=1}^{2} \beta_j = 0$$

$$\sum_{j=1}^{2} (\tau\beta)_{ij} = 0 \qquad \text{for each } j$$

$$\sum_{j=1}^{2} (\tau\beta)_{ij} = 0 \qquad \text{for each } i$$

ε_{ijk} are independent, and normally distributed with mean zero and variance σ^2.

■ *Example 4.2: A 2^2 Model*

As part of a series of experiments on the lithography develop process, a 2^2 factorial experiment was run to investigate the effects of developer dispense amount (DISP, in mL) and spin speed (SPIN, in revolutions per minute) on electrically measured critical dimension (CD, in microns (μm)). The experimental medium was polysilicon defect monitor wafers.

Each experimental factor was varied to two settings, and three wafers were processed at each of the four possible combinations. The experiment layout is shown below:

	SPIN	
DISP	3000	3300
24	3 wafers	3 wafers
27	3 wafers	3 wafers

The model that describes this experiment is:

$$y_{ijk} = \mu + \tau_i + \beta_j + (\tau\beta)_{ij} + \varepsilon_{ijk}$$

where y_{ijk} is the CD at the kth replicate of the ith level of DISP (dispense amount) and the jth level of SPIN (spin speed).

ε_{ijk} is the error from the model for this observation. Both i and j range from 1 to 2, and k ranges from 1 to 3.

4.4.1 Effects Estimates

Because each factor has only two levels, the effect of a factor can be summarized as the difference between the effect of the response its high level minus its effect at its low level:

$$\text{Effect} = \tau_2 - \tau_1$$

This effect is estimated by finding the average of all observations at the high level of the factor, then subtracting the average of all observations at the low level of the factor.

Note that this is different from the definition of "effect" for simple comparative experiments; in that case, an effect was estimated for each level of the factor, but now the entire effect of a factor is summarized in one number.

■ *Example 4.3: Estimating a Factor Effect*

Raw data from the 2^2 lithography develop factorial is shown below in standard order. The treatment combinations were actually run in random order:

DISP	SPIN	CD		
24	3000	3.51	3.97	3.61
24	3300	3.63	3.73	3.58
27	3000	3.78	3.66	3.85
27	3300	3.02	3.29	3.10

The effect of the DISP factor was estimated by subtracting the average CD at the low value (24, coded −1 later) from the average CD at the high value (27, coded 1 later). The estimated DISP effect is −0.2217.

DISP	Mean
24	3.6717
27	3.4500
Overall	3.5608

The effect of SPIN was estimated in the same manner; the estimated effect is −0.3383.

SPIN	Mean
3000	3.7300
3300	3.3917
Overall	3.5608

Interaction effects are also summarized to a single number for each type of interaction. Factor levels are coded to −1 for the lower level and +1 for the upper level, and these levels are multiplied to arrive at a similar coding for each combination of factor effects:

Factor 1	Factor 2	Interaction
−1	−1	1
−1	1	−1
1	−1	−1
1	1	1

The effect of an interaction is then defined to be the average response at the high value of the interaction minus the average response at the low value of the interaction.

■ *Example 4.4: Estimating an Interaction Effect*

For the data from the 2^2 develop step experiment, cell means were coded according to their value for the DISP*SPIN interaction as shown below. The interaction plots (Figure 4.6) based on these cell means seem to indicate some interaction.

DISP	Coded DISP	SPIN	Coded SPIN	Coded Interaction	+1 Cell Means	−1 Cell Means
24	−1	3000	−1	1	3.6967	
24	−1	3300	1	−1		3.6467
27	1	3000	−1	−1		3.7633
27	1	3300	1	1	3.1367	
		Average of cell means			3.4167	3.7050

The interaction effect was estimated by finding the difference between average cell means for the high and low values of the interaction:

$$3.4167 - 3.7050 = -0.2883$$

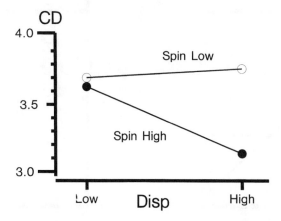

Fig. 4.6   Cell mean CDs are plotted in this interaction plot. The two values for DISP are shown on the *x*-axis, and separate lines connect means for the low (open circle) and high (solid circle) values of SPIN. These lines are not parallel, so the factors interact.

This summarized interaction effect concisely quantifies the lack of parallelism in an interaction plot: the interaction effect can only be zero if the lines are exactly parallel, and large deviations in parallelism result in large interaction effects.

4.4.2 Interpreting Interactions

The presence of a significant interaction complicates the interpretation of the effects of the individual factors, because the factors are no longer accurate predictors when considered one at a time. In a develop experiment, the effect of a change in spin speed cannot be predicted unless the setting for dispense amount is already known.

The estimate of the interaction effect itself lends little aid to the interpretation of the effect, since the same interaction estimate can be obtained in a variety of different physical situations. Consider Figure 4.7, in which four different interaction plots are shown. These all represent interactions of the same magnitude—the size

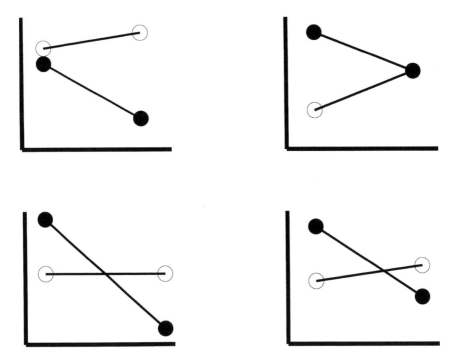

Fig. 4.7   These four hypothetical interaction plots all have the same value for the interaction effect, but they differ in the nature of the interaction they indicate. The top left graph shows two factors that have similar responses at the lower level of the first factor, but different values at the higher level. The remaining graphs show entirely different patterns of effect.

of the interaction effect is the same—but the nature is clearly different from one plot to the next.

If interaction effects are large, then it may be best to treat the factor level combinations as the (four) levels of a synthetic qualitative factor. A reanalysis of the data with this single factor may shed light on the real relationship between the original factors and the response.

In the tea brewing example, the four different treatment combinations produce only three distinct results: the 360-minute brews achieve the same strength, so a new factor with three levels might be easier to interpret:

New Factor Level	Temperature	Time
1	45	4
2	45	90
3	Any	360

In some cases, knowledge about applicable physics or chemistry can be exploited to simplify matters. If, for example, experimental factors were the length and width of a capacitor, and capacitance were the response, then the product of the two factors would probably explain the influence of both individual factors. (Capacitance is proportional to the area of the capacitor.)

Sometimes a more direct predictor is suggested by the original variables. In the travel time example, the most important predictor of driving time is the amount of traffic on the road at the time of travel. Using this factor rather than time of day and direction would produce simpler and more reliable predictions of travel time.

In many real experimental contexts, none of these simplifying options are effective, so more complex models will have to be applied—response surface models, for example.

4.4.3 Assessing Statistical Significance

Hypothesis tests are used to determine the significance of each factor or interaction effect. The hypotheses tested are:

$$H_0: \tau_1 = \tau_2 = 0$$
$$H_A: \tau_1 \neq 0 \quad \text{or} \quad \tau_2 \neq 0$$

The test is performed by comparing the sample F-statistic with the appropriate F-distribution percentile; reject H_0 if and only if:

$$\frac{\text{MST}}{\text{MSE}} > F_{1,\text{DFE}, 0.95}$$

where MST is the mean square due to the treatment (the factor or interaction) being tested, MSE is the mean square error, and DFE is the degrees of freedom for error.

MST is especially easy to find for 2^k factorial experiments replicated n times:

$$\text{MST} = n2^{k-2}\,(\text{Effect})^2$$

where Effect is the effect estimate—the difference between the average response at the high and low values of the factor (or interaction).

MSE is the estimate of within-group variance obtained by pooling estimates from every cell of the experiment; this will be demonstrated in the example that follows.

The DFE is given by:

$$\text{DFE} = 2^k(n-1)$$

■ *Example 4.5: Analysis of a Factorial Experiment*

Estimates of the main effects and the simple interaction have already been obtained in the 2^2 lithography develop factorial. Square these effects to get the MST for each of these terms and multiply it by 3 (2^{p-2}), which is 3:

Term	Effect	MST
DISP	−0.2217	0.1475
SPIN	−0.3383	0.3433
SPIN*DISP	−0.2883	0.2494

MSE is the estimate of within-group variance obtained by pooling estimates from every cell of the experiment: find the variances within each treatment combination, and average them:

DISP	SPIN	CD Variance
24	3000	0.05853
24	3300	0.00583
27	3000	0.00923
27	3300	0.01923
	Average	0.02321

The hypothesis tests can now be completed by comparing each F-ratio (MST/MSE) with $F_{1,8,0.95} = 5.32$:

Term	Effect	MST	F
DISP	−0.2217	0.1475	6.36
SPIN	−0.3383	0.3433	14.79
SPIN*DISP	−0.2883	0.2494	10.75

According to this test, all three effects are statistically significant because they are greater than 5.32.

Any statistical software will do these same tests. Minitab is one such program, and output from Minitab is shown for this test below.

■ *Example 4.6: Minitab Analysis of a Factorial Experiment*

The 2^2 factorial data was analyzed with Minitab with the following result:

Analysis of Variance for Critical Dimension

Source	DF	Seq SS	Adj SS	Adj MS	F	P
DISP	1	0.14741	0.14741	0.14741	6.35	0.036
Spin	1	0.34341	0.34341	0.34341	14.80	0.005
DISP*Spin	1	0.24941	0.24941	0.24941	10.75	0.011
Error	8	0.18567	0.18567	0.02321		
Total	11	0.92589				

Abbreviations: Adj MS, adjusted mean squares; Adj SS, adjusted sum of squares; DF, degrees of freedom;   F, sample F-value; P, probability of observing a sample F-value this large under H_0; Seq SS, sequential sum of squares.

The results obtained are the same as those obtained by the manual computation method. *P*-values show the probability of observing an *F*-value as large as the one observed. If false alarm risk (alpha risk, or α) is 0.05, then the null hypothesis should be rejected whenever the p-value is less than 0.05.

4.4.4 Checking Assumptions

The analysis method used for factorial experiments has exactly the same theoretical underpinnings that make ANOVA work in simple comparative experiments, so the same assumptions apply.

Residual analysis is the primary tool used for assumption checking. Recall that a residual is the difference between the response actually observed and the response that is predicted by the model. In full factorial experiments—those in which all possible interactions are included in the model—the predicted value for a term is just the *cell average.*

■ *Example 4.7: Residual Analysis of a Factorial Experiment*

DISP	SPIN	CD	Mean CD	Residual
24	3000	3.51	3.69667	−0.18667
24	3000	3.97	3.69667	0.27333
24	3000	3.61	3.69667	−0.08667
24	3300	3.63	3.64667	−0.01667
24	3300	3.73	3.64667	0.08333
24	3300	3.58	3.64667	−0.06667
27	3000	3.78	3.76333	0.01667
27	3000	3.66	3.76333	−0.10333
27	3000	3.85	3.76333	0.08667
27	3300	3.02	3.13667	−0.11667
27	3300	3.29	3.13667	0.15333
27	3300	3.10	3.13667	−0.03667

- A histogram of the residuals reveals no obvious departure from normality (Figure 4.8).
- A plot of residuals over the experimental factor combinations (Figure 4.9) shows they have about the same variance about zero. (Factor combinations were coded 1 for Low and 2 for High in this graph.) Outliers are the most common source of problems in satisfying the homogeneity of variance assumption, and they would have been easily spotted on this plot.

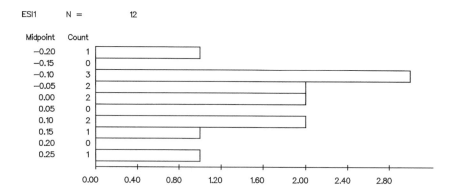

Fig. 4.8   A histogram of the residuals from a 2^2 factorial experiment appears reasonably normal.

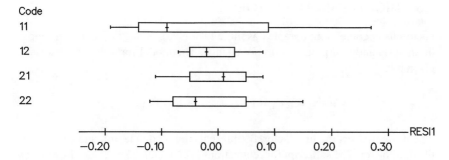

Fig. 4.9   A plot of residuals over the experiment factor combinations shows they have about the same variance about zero. Factor combinations were coded 1 for Low and 2 for High in this graph, so 12 means that DISP took its lower setting and SPIN took its higher setting.

- A plot the residuals in time order (Figure 4.10) exhibits no evidence of autocorrelation or trends.

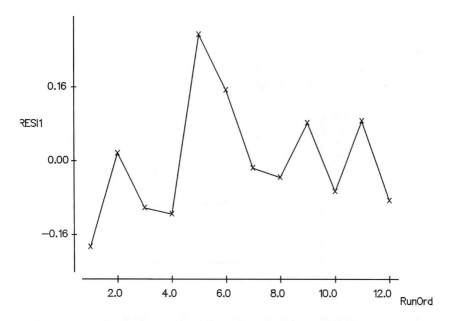

Fig. 4.10   Residuals from a 2^2 factorial experiment are plotted in time order. Neither trends nor autocorrelation (correlation between successive measurements) are apparent.

4.5 HIGHER-ORDER FACTORIALS

Factorials become more complex as more factors are involved, but the principles of analysis and interpretation do not change. The model for a 2^3 full factorial is given by:

$$y_{ijkm} = \mu + \tau_i + \beta_j + \gamma_k + (\tau\beta)_{ij} + (\tau\gamma)_{ik} + (\beta\gamma)_{jk} + (\tau\beta\gamma)_{ijk} + \varepsilon_{ijkm}$$

where τ represents the first factor, β represents the second factor, and γ represents the third factor. The subscript for replications is *m,* which takes values from 1 to *n.*

The most novel characteristic of this model is the *higher-order* (three-factor) interaction terms, the $(\tau\beta\gamma)_{ijk}$, in addition to the *simple* (two-factor) interactions. These higher-order interaction terms account for effects which are due to the simultaneous combination of several factors, but which cannot be explained by main factor or simple interaction terms.

In the driving-time example the time of day and the direction driven were both significant predictors of driving time, as was their interaction. Suppose that a third factor were introduced into this model which took the value –1 on working days (as shown in Figure 4.2), and +1 on holidays and weekends. Figure 4.11 shows the interaction plot for the +1 half of this factorial. The nature of the interaction is dif-

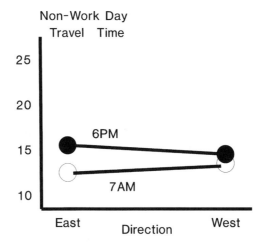

Fig. 4.11   An entirely different pattern of interaction between direction and time of day is seen on nonwork days.

ferent than it is in the −1 half of the factorial, so the three-factor interaction is important here.

Higher-order interactions occur rarely in real processes, and if they are present, their magnitude is generally much smaller than that of main effects or simple interactions.

A 2^4 develop process factorial is used in the example below to illustrate the analysis and interpretation of higher-order factorials.

■ *Example 4.8: A 2^4 Factorial*

As part of the develop process characterization, a 2^4 factorial experiment was run. Indicative responses were wafer average CD, and wafer standard deviation CD estimated from the electrical resistance of 60 polysilicon serpentines on each wafer. Economic responses were wafer yield and resist profile. Resist profile was monitored with a SEM on one unetched sacrificial wafer per treatment combination—the wafer is sacrificial because it must be broken to fit into the SEM.

Experimental factors were:

- Postexposure bake (PEB)—taking two levels: "−" indicates no bake was done, "+" indicates the bake was done. This is a categorical factor.
- Developer concentration (CONC)—taking two levels: 2.85 and 2.95 mol.
- Develop time (DTIME), taking the levels 30 seconds and 35 seconds.
- Rinse time (RTIME), taking the levels 15 seconds and 30 seconds.

One replication of the experiment takes 16 wafers, and only two replicates were done. The raw data is shown on the next page.

Observations are shown in standard order, but the actual run order was randomized—the RUNORD variable has the actual run order.

Analysis was performed as before for each continuous response; Minitab output is shown below: In this analysis of variance table, "Source" indicates the source of the variance—a factor, interaction, or error term. "DF" stands for the degrees of freedom associated with the source of variance. "Seq SS" is the sequential sum of squares—in this type of model this is the same as the sum of squares as computed by hand. "Adj SS" is the sum of squares computed as if the factor listed were the last one entered into the model, and in this type of experiment, this is the same as the sequential sum of squares. The "Adj MS" is the adjusted mean square—the sum of squares divided by the degrees of freedom. "F" is the sample F-value obtained by dividing the adjusted mean squares by the error mean square, and "P" is the p-value associated with this F-value".

PEB	CONC	DTIME	RTIME	RUNORD	CD	CDSD	YIELD
−1	−1	−1	−1	9	4.05	0.37	510
−1	−1	−1	−1	21	4.91	0.34	573
−1	−1	−1	1	11	4.61	0.29	528
−1	−1	−1	1	30	3.99	0.35	542
−1	−1	1	−1	12	4.26	0.39	548
−1	−1	1	−1	16	3.82	0.32	537
−1	−1	1	1	1	4.17	0.37	550
−1	−1	1	1	25	3.88	0.38	543
−1	1	−1	−1	14	3.67	0.33	535
−1	1	−1	−1	17	3.85	0.40	545
−1	1	−1	1	18	3.71	0.40	539
−1	1	−1	1	27	3.84	0.44	557
−1	1	1	−1	2	2.88	0.32	558
−1	1	1	−1	32	2.82	0.28	583
−1	1	1	1	15	3.41	0.32	549
−1	1	1	1	28	3.19	0.41	564
1	−1	−1	−1	4	3.36	0.14	543
1	−1	−1	−1	6	3.36	0.10	552
1	−1	−1	1	19	3.10	0.24	549
1	−1	−1	1	22	3.50	0.13	538
1	−1	1	−1	7	2.78	0.35	539
1	−1	1	−1	29	3.04	0.27	569
1	−1	1	1	24	3.32	0.16	545
1	−1	1	1	31	3.12	0.20	561
1	1	−1	−1	3	3.27	0.00	546
1	1	−1	−1	13	3.06	0.24	508
1	1	−1	1	8	3.46	0.28	543
1	1	−1	1	26	3.60	0.22	562
1	1	1	−1	5	3.46	0.30	533
1	1	1	−1	10	3.59	0.11	547
1	1	1	1	20	3.64	0.24	544
1	1	1	1	23	3.14	0.16	547

Abbreviations: CDSD, critical dimension standard deviation.

For wafer average CD, statistically significant effects were PEB, CONC, DTIME, PEB*CONC, PEB*DTIME, and (nearly) PEB*CONC*DTIME. That last term is a *three-way interaction*. The interaction plot (Figure 4.12) shows that the interaction of DTIME and CONC is seems to be affected by the level of PEB, hence the importance of PEB*CONC*DTIME.

Analysis of Variance for CD

Source	DF	Seq SS	Adj SS	Adj MS	F	P
Peb	1	2.13211	2.13211	2.13211	32.47	0.000
Conc	1	0.68445	0.68445	0.68445	10.42	0.005
dtime	1	0.72601	0.72601	0.72601	11.06	0.004
Rtime	1	0.07031	0.07031	0.07031	1.07	0.316
Peb*Conc	1	1.98005	1.98005	1.98005	30.15	0.000
Peb*dtime	1	0.40051	0.40051	0.40051	6.10	0.025
Peb*Rtime	1	0.00551	0.00551	0.00551	0.08	0.776
Conc*dtime	1	0.00080	0.00080	0.00080	0.01	0.913
Conc*Rtime	1	0.05120	0.05120	0.05120	0.78	0.390
Dtime*Rtime	1	0.02761	0.02761	0.02761	0.42	0.526
Peb*Conc*dtime	1	0.25205	0.25205	0.25205	3.84	0.068
Peb*Conc*Rtime	1	0.05780	0.05780	0.05780	0.88	0.362
Peb*dtime*Rtime	1	0.06661	0.06661	0.06661	1.01	0.329
Conc*dtime*Rtime	1	0.04500	0.04500	0.04500	0.69	0.420
Peb*Conc*dtime*Rtime	1	0.16245	0.16245	0.16245	2.47	0.135
Error	16	1.05060	1.05060	0.06566		
Total	31	7.71309				

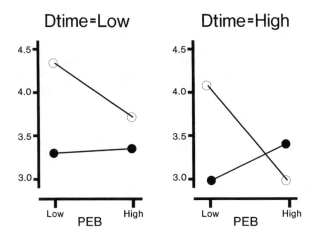

Fig. 4.12   These interaction plots seem to indicate a three-factor interaction between develop time (DTIME), developer concentration (CONC), and the presence or absence of the postexposure bake (PEB). The graph on the left shows the DTIME*PEB interaction for low concentration; the graph on the right shows the interaction for high DTIME. These shape of these plots differ, so there is a third-order interaction.

A histogram of residuals (Figure 4.13) does not reveal any radical departures from normality, and a plot by group for the first two significant factors (Figure 4.14) reveals no serious violations of the equality of variance assumption. (An equally informative plot could have been made using other factors.)

Finally, a plot of residuals in time order (Figure 4.15) seems to indicate that the experiment was not adversely affected by confounding.

The analysis of the second indicative response, CD standard deviation (CDSD), was simpler:

Analysis of Variance for CDSD						
Source	DF	Seq SS	Adj SS	Adj MS	F	P
Peb	1	0.206403	0.206403	0.206403	43.71	0.000
Conc	1	0.000078	0.000078	0.000078	0.02	0.899
dtime	1	0.003003	0.003003	0.003003	0.64	0.437
Rtime	1	0.003403	0.003403	0.003403	0.72	0.408
Peb*Conc	1	0.000528	0.000528	0.000528	0.11	0.742
Peb*dtime	1	0.010153	0.010153	0.010153	2.15	0.162
Peb*Rtime	1	0.000253	0.000253	0.000253	0.05	0.820
Conc*dtime	1	0.013203	0.013203	0.013203	2.80	0.114
Conc*Rtime	1	0.013203	0.013203	0.013203	2.80	0.114
dtime*Rtime	1	0.008778	0.008778	0.008778	1.86	0.192
Peb*Conc*dtime	1	0.000078	0.000078	0.000078	0.02	0.899
Peb*Conc*Rtime	1	0.000378	0.000378	0.000378	0.08	0.781
Peb*dtime*Rtime	1	0.019503	0.019503	0.019503	4.13	0.059
Conc*dtime*Rtime	1	0.000028	0.000028	0.000028	0.01	0.939
Peb*Conc*dtime*Rtime	1	0.001378	0.001378	0.001378	0.29	0.596
Error	16	0.075550	0.075550	0.004722		
Total	31	0.355922				

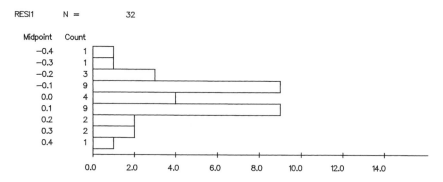

RESI1   N =   32

Midpoint	Count	
−0.4	1	
−0.3	1	
−0.2	3	
−0.1	9	
0.0	4	
0.1	9	
0.2	2	
0.3	2	
0.4	1	

0.0   2.0   4.0   6.0   8.0   10.0   12.0   14.0

Fig. 4.13   A histogram of residuals for the CD response of the 2^4 factorial does not reveal any obvious departures from normality.

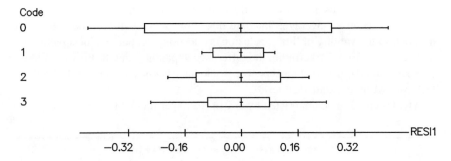

Fig. 4.14   A group of CD residual boxplots by the first two statistically significant factors of the 2^2 factorial is shown. Factor combinations were coded: $0 = (-1, -1)$, $1 = (-1, 1)$, $2 = (1, -1)$, and $3 = (1, 1)$. The variance within each treatment combination is not exactly the same, but also not so different that the model assumptions seem to have been violated.

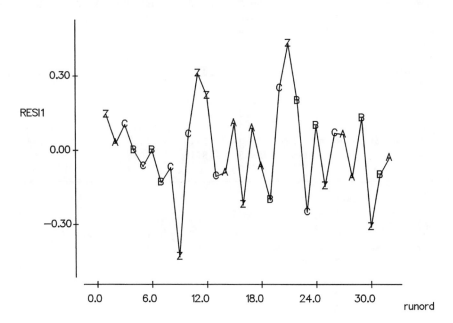

Fig. 4.15   A plot of CD residuals in time order seems to indicate that the experiment was not adversely affected by confounding.

PEB seems to have a very large effect on this component of variance, decreasing it by 0.16 on the average. Standard deviations are usually not normally distributed, so the validity of this conclusion is questionable, but residuals seem normal (Figure 4.16). Furthermore, the difference apparently due to PEB is so large compared to all others that most people would have been convinced of the PEB effect even without a statistical test.

Yield seemed to be unaffected by any of the experimental factors:

Analysis of Variance for Yield

Source	DF	Seq SS	Adj SS	Adj MS	F	P
Peb	1	38.3	38.3	38.3	0.14	0.717
Conc	1	34.0	34.0	34.0	0.12	0.732
dtime	1	675.3	675.3	675.3	2.41	0.140
Rtime	1	38.3	38.3	38.3	0.14	0.717
Peb*Conc	1	850.8	850.8	850.8	3.03	0.101
Peb*dtime	1	108.8	108.8	108.8	0.39	0.542
Peb*Rtime	1	148.8	148.8	148.8	0.53	0.477
Conc*dtime	1	34.0	34.0	34.0	0.12	0.732
Conc*Rtime	1	132.0	132.0	132.0	0.47	0.503
dtime*Rtime	1	101.5	101.5	101.5	0.36	0.556
Peb*Conc*dtime	1	166.5	166.5	166.5	0.59	0.452
Peb*Conc*Rtime	1	195.0	195.0	195.0	0.70	0.417
Peb*dtime*Rtime	1	3.8	3.8	3.8	0.01	0.909
Conc*dtime*Rtime	1	385.0	385.0	385.0	1.37	0.259
Peb*Conc*dtime*Rtime	1	11.3	11.3	11.3	0.04	0.844
Error	16	4488.5	4488.5	280.5		
Total	31	7412.0				

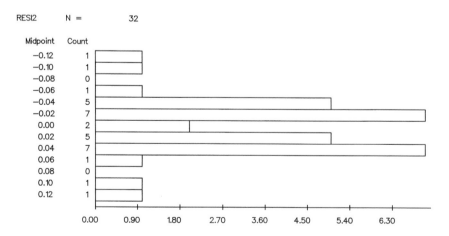

Fig. 4.16 Residuals for the CD standard deviation seem to be normally distributed.

Results of the second economic response were less encouraging: resist profiles (Figure 4.17) for the entire PEB half of the experiment exhibited obvious resist flow. Although this is a subjective judgement based on qualitative measurements, it still provides compelling evidence that this particular PEB recipe is unacceptable. It also throws into question all of the CD measurements on the PEB half of the 2^4 factorial, because resist flow probably affects the following etch in some way.

In spite of the fact that half of the response measurements in the example above might have been wrong, the experiment was not a total waste. The non-PEB half of the experiment is a 2^3 factorial which could yield some valuable information on the effects of DTIME, RTIME, and CONC. This ability of factorials to "collapse" into smaller factorials is one of their most attractive properties.

4.6 UNREPLICATED FACTORIALS

Replication makes the process of experimentation more robust, because even if some material is lost useful results are still obtained. Replication also provides a convenient measure of within-group variance which can be used to test the statistical significance of observed effects. But replication is expensive: material, time, and labor costs are all higher in replicated experiments than in unreplicated experiments. This section presents some statistical techniques that can be applied when the costs of replication are not justified by its benefits.

Unreplicated experiments can be divided into two classes: those that have some partial replication (some treatment combinations are repeated), and those in which no treatment combination is repeated in the entire experiment. The most common

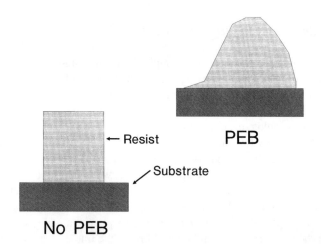

Fig. 4.17   These cross-sectional SEMS of exposed and developed wafers show that the resist profile for the PEB side of the experiment is unacceptable.

approach in the first case is to add centerpoints to the experiment. Analysis of this type of experiment is the subject of Section 4.6.1. Completely unreplicated experiments are covered in Section 4.6.2.

4.6.1 Unreplicated Factorials With Centerpoints

Planning an unreplicated experiment with centerpoints is rather easy since the main issue of sample size has already been resolved. Statistical tests conducted on data from unreplicated factorials will be relatively insensitive—they will have higher beta risk so more real differences will go undetected than in replicated experiments.

Beta risk can be controlled to some degree by the number of centerpoints included in the experiment. At the very least, three centerpoints should be used: one centerpoint does not give any measure of within-group variance, and two centerpoints give a very poor estimate. Including more than a few centerpoints is not an effective design technique—the fractional replication methods presented in Chapter 5 are a better choice.

Randomization is even more critical in unreplicated factorials, because it is much more difficult to detect confounding than before. A plot of residuals in time order can usually uncover confounding, but in an unreplicated factorial residuals are zero by definition except at the centerpoints. Centerpoints must be randomized along with the other experimental runs, as they are the only indicator of confounding.

Analysis of unreplicated factorials with centerpoints is a little more difficult than before: the experiment is now unbalanced, so statistical packages must be utilized carefully to produce correct answers. The procedure is illustrated in the example below.

■ *Example 4.9: An Unreplicated Factorial with Centerpoints*

Suppose that in the experiment of Example 4.8 centerpoints were run instead. A SAS analysis of data from that experiment finds PEB and DTIME to have significant effects on CD; several effects that previously tested significant are now found to be insignificant. Pr > F is the p-value associated with the F-test.

This analysis yields meaningful results for Type III SS, but tests using Type I SS accord deceptively low influence to PEB. This would happen to any factor that happened to be listed first in the SAS *Model* statement. This is not a flaw in the software—all results produced are technically correct when interpreted appropriately.

The only residuals available in this type of experiment are those at the centerpoints, so assumption checking is nearly impossible. Centerpoints can be used to check linearity: plot the actual data observed by any variable of interest—responses at the centerpoints should lie roughly on the line determined the "–" and the "+" endpoints.

		Dependent Variable: CD		
Source	DF	Sum of Squares	Mean Square	Pr > F
Model	16	3.95777500	0.24736094	0.1616
Error	3	0.20840000	0.06946667	
Corrected total	19	4.16617500		
Source	DF	Type III SS	Mean Square	Pr > F
PEB	1	0.94575625	0.94575625	0.0345
CONC	1	0.40005625	0.40005625	0.0959
PEB*CONC	1	0.66830625	0.66830625	0.0532
DTIME	1	0.77000625	0.77000625	0.0447
PEB*DTIME	1	0.31640625	0.31640625	0.1225
CONC*DTIME	1	0.00525625	0.00525625	0.8011
PEB*CONC*DTIME	1	0.09765625	0.09765625	0.3211
RTIME	1	0.00225625	0.00225625	0.8685
PEB*RTIME	1	0.04100625	0.04100625	0.4982
CONC*RTIME	1	0.07425625	0.07425625	0.3772
PEB*CONC*RTIME	1	0.11390625	0.11390625	0.2904
DTIME*RTIME	1	0.00600625	0.00600625	0.7879
PEB*DTIME*RTIME	1	0.36300625	0.36300625	0.1064
CONC*DTIME*RTIME	1	0.14630625	0.14630625	0.2426
PEB*CONC*DTIME*RTIME	1	0.00680625	0.00680625	0.7748

4.6.2 Unreplicated Factorials Without Centerpoints

Analysis of factorials without centerpoints (or some other intrinsic measure of within-group variance) can be challenging. Statistical software packages will still produce sums of squares as before, but it is impossible to actually compute an *F*-statistic, so entirely objective statistical tests are not available. Two popular solutions to this problem are to use normal probability plots, or collapse the factorial over seemingly insignificant factors and interactions.

Normal probability plots, or probplots for short, use the ANOVA assumptions to arrive at a visual "test" of the significance of factor effects. According to those assumptions, if all factor effects are truly zero, then their estimates will be normally distributed with a mean of zero. By plotting estimated effects on graph paper on which one axis is scaled by normal distribution probability, significant effects should be apparent because they will not fall on the line determined the rest of the effects, or they will fall far from the center of the plot.

■ *Example 4.10: Normal Probability Plot Analysis of an Unreplicated Factorial*

Effects estimates from the unreplicated 2^4 experiment above were computed with the results shown below:

Term	Effect
RTM	−0.02
DTM	−0.44
DTM*RTM	0.04
CON	−0.32
CON*RTM	0.14
CON*DTM	0.04
CON*DTM*RTM	−0.19
PEB	−0.49
PEB*RTM	0.10
PEB*DTM	0.28
PEB*DTM*RTM	−0.30
PEB*CONC	0.41
PEB*CONC*RTM	−0.17
PEB*CONC*DTM	0.16
PEB*CONC*DTM*RTM	−0.04

A normal probability plot of these effects estimates is shown in Figure 4.18. PEB and DTIME (the two lowest points) deviate a bit from the line through the rest of the points to appear significant, and PEB*CONC (highest point) may also be significant.

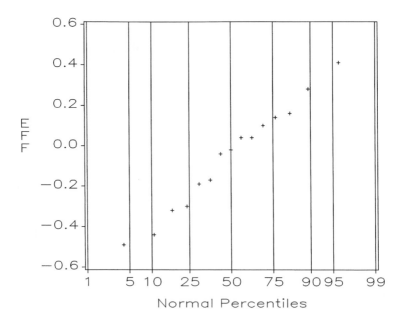

Fig. 4.18   This normal probability plot of effects estimates can be used to visually assess effect significance. Effects that are far from the 50th percentile are suspect, as are those that deviate far from the line through the center of the plotted points. The two smallest effects and the largest one definitely fall into this category.

Probability plot analysis is a bit subjective; with just a little more imagination the experimenter in the example above could have found the PEB*DTIME interaction to be significant. Somewhat more objective methods of interpretation have been suggested—see Lenth (1989) for example.

One benefit of the probability plot is that it points out the *non*significance of some factors or interactions. This discovery can simplify the analysis by suggesting that these apparently unimportant terms be left out of the model, thus furnishing replicates which can be used to estimate error.

This approach is not without risk, and could lead the experimenter to do one of the following:

- Commit a Type II error by deciding that some factors had no effect when they really did.
- Enhance beta risk for remaining factors by inflating variance estimates used to test their significance.

A safer approach is to assume before the analysis commences that interactions of some given order and above are not significant, and use their sums of squares as an estimate of error.

■ *Example 4.11: Reanalysis of an Unreplicated Factorial Without Higher-Order Interactions*

A second analysis of the unreplicated 2^4 factorial without centerpoints was done; this time all three-way and higher interactions were left out of the model—there are five of these. If these effects are truly not significant, their estimates are normally distributed with zero mean and variance equal to the within-group variance. A SAS analysis using this assumption is shows the PEB is a significant factor:

General Linear Models Procedure—Dependent Variable: CD				
Source	DE	Sum of Squares	Mean Square	Pr > F
Model	10	3.22931250	0.32293125	0.1961
Error	5	0.72768125	0.14553625	
Corrected total	15	3.95699375		

Source	DF	Type III SS	Mean Square	Pr > F
PEB	1	0.94575625	0.94575625	0.0513
CONC	1	0.40005625	0.40005625	0.1582
DTIME	1	0.77000625	0.77000625	0.0698
RTIME	1	0.00225625	0.00225625	0.9058
PEB*CONC	1	0.66830625	0.66830625	0.0850
PEB*DTIME	1	0.31640625	0.31640625	0.2004
PEB*RTIME	1	0.04100625	0.04100625	0.6183
CONC*DTIME	1	0.00525625	0.00525625	0.8567
CONC*RTIME	1	0.07425625	0.07425625	0.5070
DTIME*RTIME	1	0.00600625	0.00600625	0.8470

4.7 FACTORIALS AND BLOCKING

Blocking factors like lot and shift affect factorial experiments just as they do simpler comparative experiments, so it should be no surprise that blocking is a very effective money saver in factorial experiments. The analysis of blocked factorials is essentially the same as before; a presentation of this standard analysis will be found in Section 4.7.1. Planning and analyzing a blocked factorial experiment requires more care if the factorial is too large to fit into a single block. The scheme of deliberate confounding used in this situation is presented in Section 4.7.2.

4.7.1 Blocks Containing Factorials

If a factorial can be entirely contained within natural blocks, analysis proceeds just as before except that the block effects must be accounted for.

■ *Example 4.12: A Blocked Factorial Experiment*

The focus of the third develop experiment was to find a PEB recipe that still produced excellent cross-wafer uniformity (small CD standard deviation), but did not cause the resist to flow. Since SEM resist flow measurements are costly and time consuming, this experiment was done on six-wafer lots of a product which had an electrical structure designed to measure contact resistance. This structure is a chain of contacts (Figure 4.19) that is particularly sensitive to changes in resist profiles, and had been used in previous experiments to indirectly gauge resist flow so it was considered a good surrogate for the SEM previously used to evaluate resist flow.

Product lot is known to be an important blocking factor, so a single replicate of a 2^2 factorial experiment was performed within each of the three experimental lots. The two experimental factors were postexposure bake time and temperature; all

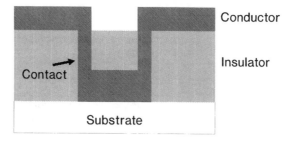

Fig. 4.19   This is one link in a contact chain that can be used to assess resist profiles. Changes in resist profile will cause the edges of the etched contact to change; the thickness of the conductor into the contact will change because of this, and the resistance of the entire chain will reflect that change.

other factors were held at target values. The remaining two wafers in each lot were centerpoints so that the linearity assumption of the model could be tested.

A centerpoint would normally be a value run at the process target settings, but PEB was not previously part of this process so a new "center" was selected. Since the PEB recipe used in the last experiment was entirely unacceptable due to resist flow, that recipe could not be used for centerpoints. Instead, a bit of engineering judgment was applied to move the entire experiment away from the most likely cause of excessive resist flow—high temperature. The previous PEB recipe was 45 seconds at 120°C; the centerpoints for this new experiment were at 45 seconds at 80°C.

Raw data from the experiment is shown below. On the advice of the Electrical Test engineer, a log transform (LCR) was used on the contact resistance measurements to normalize residuals. This transform is often successful in normalizing data that vary over a number of orders of magnitude.

LOT	TIME	TEMP	CR	LCR
1	30	60	27	3.29584
2	30	60	256	5.54518
3	30	60	128	4.85203
1	30	100	451	6.11147
2	30	100	418	6.03548
3	30	100	992	6.89972
1	45	80	165	5.10595
1	45	80	520	6.25383
2	45	80	120	4.78749
2	45	80	697	6.54679
3	45	80	271	5.60212
3	45	80	869	6.76734
1	60	60	111	4.70953
2	60	60	1,386	7.23418
3	60	60	407	6.00881
1	60	100	1,976	7.58883
2	60	100	19,058	9.85524
3	60	100	10,138	9.22405

Abbreviations: CR, .

A SAS analysis of this table (below) reveals that both the experimental factors have a statistically significant effect on contact resistance; the interaction of time and temperature is not statistically significant.

The histogram in Figure 4.20 provides visual evidence that the log transformation of contact resistance measurements successfully normalized residuals.

General Linear Models Procedure—Dependent Variable: LCR

Source	DF	Sum of Squares	Mean Square	Pr > F
Model	6	35.54901466	5.92483578	0.0011
Error	11	7.37447812	0.67040710	
Corrected total	17	42.92349278		

Source	DF	Type III SS	Mean Square	Pr > F
LOT	2	4.89546527	2.44773264	0.0608
TIME	1	11.76302613	11.76302613	0.0015
TEMP	1	16.49525738	16.49525738	0.0004
TIME*TEMP	1	0.94190367	0.94190367	0.2609

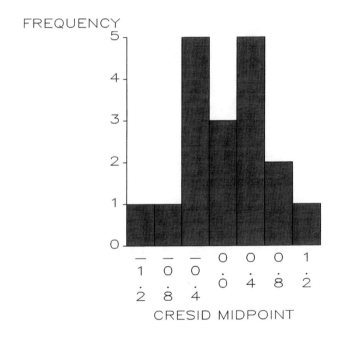

Fig. 4.20   This histogram provides visual evidence that the log transformation of contact resistance measurements successfully normalized residuals.

One reason for including centerpoints was to check the linearity of the effects. If the effects of time and temperature are truly linear, then a plot of the response averages over the factor settings should show three points all lying approximately on the same line. Figure 4.21 seems to indicate that neither the effect of time nor temperature is linear—this model does not fit.

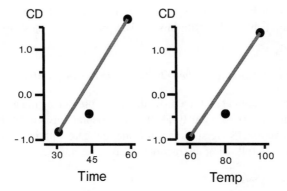

Fig. 4.21   These plots of CD by time, and by temperature, give convincing evidence that neither have a perfectly linear effect on CD.

4.7.2 Factorials Split Across Blocks

Some factorial experiments are too large to fit into a block, so some way must be found to spread factorial runs across blocks while still obtaining estimates of important effects. The design used in this situation deliberately confounds higher-order interaction with blocks, which still allows for the estimation of main effects and lower-order interactions.

Consider a very simple example in which a 2^3 factorial experiment must be run in two blocks of size 4. The design chosen has the following layout:

X1	X2	X3	X1*X2	X1*X3	X2*X3	X1*X2*X3	Block
−1	−1	−1	1	1	1	−1	A
−1	−1	1	1	−1	−1	1	B
−1	1	−1	−1	1	−1	1	B
−1	1	1	−1	−1	1	−1	A
1	−1	−1	−1	−1	1	1	B
1	−1	1	−1	1	−1	−1	A
1	1	−1	1	−1	−1	−1	A
1	1	1	1	1	1	1	B

The two blocks in the experiment are denoted A and B, and the highest-order interaction—X1*X2*X3—is deliberately confounded with blocks: in one block it takes the value 1, in the other block it takes the value −1. This will make it impossible to estimate this interaction, but every other effect can be estimated because both of their levels appear within each block.

The analysis of this type of experiment requires that the experimenter know which effects are confounded with blocks. The procedure is illustrated in the example below.

■ *Example 4.13: Confounding Higher-Order Interactions with Blocks*

Suppose that the unreplicated 2^4 factorial described previously must be performed in blocks of eight wafers. Obviously, all 16 experimental runs will not fit into a single block, but two blocks are sufficient for one replicate without centerpoints. In this experiment, the following interactions are of little interest:

PEB*CONC*DTIM
PEB*CONC*RTIME
PEB*DTIME*RTIME
CONC*DTIME*RTIME
PEB*CONC*DTIME*RTIME

The least important of these, PEB*CONC*DTIME*RTIME, will be confounded with the blocks. This means that all runs in which this interaction is −1 will be put in the first block (in random order of course), and all runs with a value of +1 for this interaction will be put in the second block. The experiment layout and results are shown below, with PCDR denoting the four-way interaction.

BLOCK	PEB	CONC	DTIME	RTIME	CD	PEB*CONC
0	−1	−1	−1	1	3.99	−1
0	−1	−1	1	−1	3.82	−1
0	−1	1	−1	−1	3.85	−1
0	−1	1	1	1	3.19	−1
0	1	−1	−1	−1	3.36	−1
0	1	−1	1	1	3.12	−1
0	1	1	−1	1	3.60	−1
0	1	1	1	−1	3.59	−1
1	−1	−1	−1	−1	4.91	1
1	−1	−1	1	1	3.88	1
1	−1	1	−1	1	3.84	1
1	−1	1	1	−1	2.82	1
1	1	−1	−1	1	3.50	1
1	1	−1	1	−1	3.04	1
1	1	1	−1	−1	3.06	1
1	1	1	1	1	3.14	1

A SAS analysis of this data for the CD response demonstrates that both main effects and simple interactions can be estimated. Three-way interactions were assumed to be zero here so that an estimate of within-group variance can be obtained.

General Linear Models Procedure—Dependent Variable: CD				
Source	DF	Type III SS	Mean Square	Pr > F
BLOCK	1	0.00680625	0.00680625	0.8554
PEB	1	0.94575625	0.94575626	0.0838
CONC	1	0.40005625	0.40005625	0.2105
DTIME	1	0.77000625	0.77000625	0.1076
RTIME	1	0.00225625	0.00225625	0.9163
PEB*CONC	1	0.66830625	0.66830625	0.1264
PEB*DTIME	1	0.31640625	0.31640625	0.2558
PEB*RTIME	1	0.04100625	0.04100625	0.6582
CONC*DTIME	1	0.00525625	0.00525625	0.8727
CONC*RTIME	1	0.07425625	0.07425625	0.5559
DTIME*RTIME	1	0.00600625	0.00600625	0.8640

There are other assignments of treatments to blocks that confound important effects with blocks, thus making it impossible to estimate these effects. One such assignment choice is shown below; the PEB*CONC interaction is confounded with blocks.

BLOCK	PEB	CONC	DTIME	RTIME	PEB*CONC
0	−1	−1	−1	−1	1
0	−1	−1	−1	1	1
0	−1	−1	1	−1	1
0	−1	−1	1	1	1
1	−1	1	−1	−1	−1
1	−1	1	−1	1	−1
1	−1	1	1	−1	−1
1	−1	1	1	1	−1
1	1	−1	−1	−1	−1
1	1	−1	−1	1	−1
1	1	−1	1	−1	−1
1	1	−1	1	1	−1
0	1	1	−1	−1	1
0	1	1	−1	1	1
0	1	1	1	−1	1
0	1	1	1	1	1

4.8 ANALYSIS OF COVARIANCE

Analysis of Covariance (ANCOVA) is an extension of linear regression analysis which involves, in its simplest form, one continuous factor and one categorical factor. Applications for ANCOVA arise naturally when data is collected from several different machines, or when an uncontrolled (but measurable) variable is observed in addition to a categorical variable.

ANCOVA would be used to analyze data from an experiment investigating the effect of developer temperature on CD, in which postexposure bake was used in part of the experiment, and not used in another part. Postexposure bake is a categorical factor—it is present, or not—and one purpose of the experiment is to test the effect of postexposure bake on CD. Temperature is a continuous factor, and another purpose of the experiment is to estimate the relationship between temperature and CD.

ANCOVA allows data about the continuous variable to be pooled over the levels of the categorical variable so that better estimates of the effect of the continuous variable are obtained. ANCOVA also allows the effects of categorical variables to be estimated and tested even if an uncontrolled continuous variable is present—as long as the continuous variable is measured, its effect on the response can be removed prior to the analysis of the effects of the categorical variable.

One model for ANCOVA is given by:

$$y_{ijk} = \mu' + \beta_1 X_i + \tau_j + \varepsilon_{ijk}$$

where:

- μ' is an overall offset (not actually the response mean).
- β_1 is the slope of the simple linear relationship between the continuous factor and the response.
- X_i is the ith value of the continuous factor.
- τ_j is the effect of the jth level of the categorical factor. The sum of these effects is zero.
- ε_{ijk} is an error from the predicted value for the kth replicate at the jth level of the categorical factor and the ith value of the continuous factor. ε_{ijk} is normally distributed with mean zero and variance σ^2.

One very important assumption in the ANCOVA model is that the relationship between the continuous factor and the response is the same at all levels of the categorical factor. This can be checked visually by plotting separate regression lines for each level of the categorical factor; these lines should have nearly the same slope and intercept. If they differ significantly, the levels of the categorical factor should be treated as separate populations. There are statistical tests for homogeneity of slope—see Neter et al. (1985). That text also presents more complex types of ANCOVA models which can accommodate different slope terms for each subpopulation.

Analysis of ANCOVA experiments is similar to that of factorial experiments, as is demonstrated in the example below.

■ *Example 4.14: Analysis of Covariance*

An experiment was done to investigate the effect of rinse time (RTIME) on CD; rinse time was varied to two levels, denoted −1 and 1. The analysis of this experiment would have been a simple ANOVA, if not for the fact that the rinse volume (RVOL) was uncontrolled. Rinse volume is a continuous variable known to affect CD, so its effects must be accounted for: ANCOVA is the most appropriate analysis method in this situation.

Raw data from the experiment is shown below:

RTIME	RVOL	CD
−1	62	3.73
−1	62	4.22
−1	65	3.70
−1	68	3.82
−1	65	4.01
−1	61	3.46
−1	70	2.85
−1	69	3.05
−1	65	3.71
1	64	2.41
1	68	2.66
1	62	3.14
1	71	2.43
1	68	2.60
1	64	3.35
1	66	2.31
1	69	2.52
1	61	3.10

A SAS analysis of the data reveals that both rinse time and rinse volume have significant effects on CD:

General Linear Models Procedure—Dependent Variable: CD			
Source	DF	Sum of Squares	PR > F
Model	2	4.64862484	0.0001
Error	15	1.57946960	
Corrected total	17	6.22809444	

Source	DF	Type III SS	Pr > F
RTIME	1	3.13994967	0.0001
RVOL	1	1.06635262	0.0062

Parameter	Estimate	T for H$_0$: Parameter=0	Pr > |T|
RVOL	−0.078184713	−3.18	0.0062

"T for H0" is the sample t-statistic for this data, and "Pr > |T|" is the p-value for the test. This p-value is interpreted in exactly the same way as a p-value for an F-test: small values (less than 0.05) indicate statistical significance.

The effect of rinse volume is indicated by the estimate of slope (–0.7818), just like in any other regression analysis.

To evaluate the assumption that the slope is the same for all levels of the categorical factor, the slope was estimated separately for the two populations of rinse time. A plot (Figure 4.22) of the raw data with superimposed regression lines seems to verify the assumption the slopes are the same in the two populations. (CDs for low rinse times are marked with squares; CDs for high rinse times are marked with triangles.)

4.9   SUMMARY

The factorial experiment is one of the great moneysavers of statistics. It allows for simultaneous experimentation with several factors, and it gives insight into how those factors interact. Knowledge of these interactions is essential in the semiconductor industry because interactions occur often, and ignoring them can lead to serious consequences.

This chapter covered those designs which account for the majority of factorial experiments—the 2^k design, perhaps blocked, perhaps with centerpoints. There are many other variations on the factorial experiment which may be useful in particular situations. Some common variations are described below:

- 3^k, $3^k \bullet 2^p$, and more complex factorials. Response surface methods shown in Chapter 6 are usually a better choice of design than these very expensive experiments.

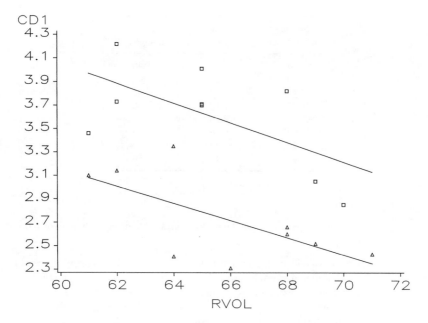

Fig. 4.22   Raw data for both rinse times was plotted: squares indicate low values for RINSE; triangles indicate high values for RINSE. Regression lines were estimated separately for these two populations, and are drawn on the plot. The lines are approximately parallel, so a simple ANCOVA model (without interaction terms) seems to fit.

- Mixture designs: a class of experiments in which factor levels are constrained such that they must add to some constant. Mixing several liquid components to form a solution of a fixed volume of a solution is one such experiment. These designs depend on different assumptions then do factorials, so the usual analysis can give misleading results. See Cornell (1981) for an exposition on this topic.

CHAPTER 4 PROBLEMS

1. A tea drinker is investigating the effects of brewing temperature and steeping time on the strength of tea. She collected the following measurements:
   (a) What type of experiment is this?
   (b) Compute point estimates for the effects of brewing temperature and steeping time.
   (c) Make an interaction plot; does there appear to be any interaction between the factors?
   (d) Do an ANOVA to test the effects of temperature, time, and their interaction.
   (e) Compute a beta risk curve for this experiment. What is the probability that this type of experiment would have detected a difference of 10 strength points?

Brewing Temperature	Brewing Time (min)	Strength (CIU)[a]
70	10	40
70	10	48
70	12	45
70	12	50
90	10	67
90	10	55
90	12	57
90	12	59

[a]CIU = color intensity units.

(*f*) How many cups of tea would have to be brewed to detect a strength difference of 5 points 90% of the time?

2. An engineer who loves pizza wants to find a combination of cooking temperature and cooking time that produces the ideal pizza: one in which the crust is crispy and light, and the toppings are cooked thoroughly but not burned. He conducted a 2^2 factorial experiment, replicated twice, and rated the pizza quality on a scale of 1 to 100 for crust quality (CRISP). Results are shown below:

TEMP	TIME	CRISP
350	10	26
350	10	38
350	12	53
350	12	60
375	10	79
375	10	60
375	12	48
375	12	51

(*a*) Compute point estimates for the effects of time and temperature on the response.

(*b*) Make interaction plots for the response. Are there any interactions that appear to be significant?

(*c*) Are any of the ANOVA assumptions violated in this experiment? What influence should this have on your interpretation of the results.

3. The pizza-loving engineer thinks that the influences of time and temperature may depend on the type of topping, so he conducted a 2^3 factorial experiment with three factors: temperature, time, and topping (-1 = pepperoni, 1 = artichoke hearts). Four

centerpoints were added to provide a measure of experimental error so statistical tests can be done. The centerpoint used pepperoni. All the treatment combinations—including centerpoints—were run in random order.

TEMP	TIME	TOP	CRISP
350	10	1	26
350	10	2	51
350	12	1	53
350	12	2	59
375	10	1	79
375	10	2	47
375	12	1	48
375	12	2	45
362	11	1	51
362	11	1	37
362	11	1	49
362	11	1	52

(*a*) Compute point estimates for the effects of time, temperature, and topping.
(*b*) Make interaction plots for the simple interactions, and for the three-way interaction. Do any interactions appear to be significant.
(*c*) Use ANOVA to test the significance of the effects and all possible interactions.
(*d*) Comment on the validity of using these particular centerpoints to provide a measure of experimental error.
(*e*) How many pizzas would be needed to detect a crust quality difference of 10 points?

4. A microlithography engineer is investigating the effects of exposure energy and focus on two responses of interest: CD and the standard deviation of the CD. Exposure energy was varied about its target of 120 mJ in either direction; focus took the values 0 [microns (μ) out of the focus plane—this is perfect focus] and 5 microns (5 μm). A 2^2 experiment was performed in which each treatment combination was replicated five times. The results are shown below in standard order (even though the actual experiment was randomized).

Exposure	Focus		CD			
110	0	3.44	3.49	3.46	3.49	3.49
110	5	3.49	3.45	3.54	3.79	3.46
130	0	3.21	3.22	3.20	3.20	3.25
130	5	3.22	3.46	3.52	3.57	3.77

(*a*) Do an analysis of variance on the CD response. Are any of the factors significant? Is their interaction significant?

(*b*) Use graphical methods to check the validity of model assumptions. Are any assumptions violated?

(*c*) Compute the standard deviation for each treatment combination, and take its square root to normalize it.

Determine point estimates for the effects of focus and energy on the standard deviation. Does either factor appear to be significant?

(*d*) Draw an interaction plot depicting the effects of focus and energy on the square root of the standard deviation.

5. The investigation of microlithography parameters was continued in another experiment which included two additional factors: resist coat thickness, and postexposure bake. A 2^4 experiment was performed with 20 wafers: 16 for the factorial, and four centerpoints. The responses were the wafer average CD, and the standard deviation of CDs on the wafer.

Exposure (mJ)	Focus (μm)	Resist Thickness (Å)	PEB[a]	CD[b] Mean	CD SD[c]
110	0	11,000	−1	3.44	0.270
110	0	11,000	1	3.46	0.018
110	0	12,000	−1	3.69	0.268
110	0	12,000	1	3.40	0.014
110	5	11,000	−1	3.79	0.312
110	5	11,000	1	3.76	0.052
110	5	12,000	−1	3.93	0.300
110	5	12,000	1	4.03	0.060
130	0	11,000	−1	3.20	0.258
130	0	11,000	1	3.22	0.022
130	0	12,000	−1	3.41	0.276
130	0	12,000	1	3.20	0.014
130	5	11,000	−1	3.34	0.306
130	5	11,000	1	3.42	0.052
130	5	12,000	−1	4.07	0.302
130	5	12,000	1	3.44	0.054
120	3	11,500	1	3.53	0.052
120	3	11,500	1	3.39	0.042
120	3	11,500	1	3.37	0.030
120	3	11,500	1	3.43	0.032

[a]PEB = post-exposure bake.
[b]CD = critical dimension.
[c]CD SD = critical dimension standard deviation.

(*a*) Use ANOVA to test the significance of the factors on wafer average CD. Are any factors or their interactions significant?

(*b*) Make interaction plots for any interactions that were significant.

(*c*) Use graphical methods to check the validity of the ANOVA assumptions.

(*d*) Repeat parts (a) through (c) using the square root of the standard deviations as a response.

(*e*) The expected loss can be used as a single response comprehending both wafer average and wafer standard deviation. Suppose that the CD target is 3.10 microns (3.10 μm), the lower specification limit is 2.85 microns (2.85 μm), and the upper specification limit is 3.35 microns (3.35 μm). If 50% of the die produced at either specification limit will fail to function, what UDNLF best describes the loss?

(*f*) Convert the original data for mean and standard deviation into expected loss, and test the significance of the factors on expected loss.

6. An additional factor has been suggested to the microlithography engineer: spin duration. He performed a 2^5 factorial including this extra factor, but was forced to split the experiment over two 16-wafer lots. The following results were obtained:

Exp.	Focus	Thick.	PEB	Spin	CD
110	0	11,000	−1	90	3.44
110	0	11,000	−1	120	3.49
110	0	11,000	1	90	3.46
110	0	11,000	1	120	3.49
110	0	12,000	−1	90	3.69
110	0	12,000	−1	120	3.61
110	0	12,000	1	90	3.40
110	0	12,000	1	120	3.42
110	5	11,000	−1	90	3.79
110	5	11,000	−1	120	3.46
110	5	11,000	1	90	3.76
110	5	11,000	1	120	3.83
110	5	12,000	−1	90	3.93
110	5	12,000	−1	120	3.92
110	5	12,000	1	90	4.03
110	5	12,000	1	120	3.47
130	0	11,000	−1	90	3.20
130	0	11,000	−1	120	3.21
130	0	11,000	1	90	3.22
130	0	11,000	1	120	3.25
130	0	12,000	−1	90	3.41
130	0	12,000	−1	120	3.45
130	0	12,000	1	90	3.20
130	0	12,000	1	120	3.17
130	5	11,000	−1	90	3.34
130	5	11,000	−1	120	3.67
130	5	11,000	1	90	3.42
130	5	11,000	1	120	3.59
130	5	12,000	−1	90	4.07
130	5	12,000	−1	120	3.55
130	5	12,000	1	90	3.44
130	5	12,000	1	120	3.34

(*a*) Why do you suppose this particular manner of splitting the factorial over the lots was used? Are there better ways to do this?

(*b*) Use ANOVA to test the significance of the main effects and their simple interactions, ignore third-order and higher interactions.

(*c*) Explain the results of the ANOVA in part (b), using interaction plots if necessary.

(*d*) What is the risk of the type of analysis used in part (b)? How could these risks have been avoided?

7. A tennis player with a rather weak serve has decided to use statistics to improve his serve speed. He conducted an unreplicated 2^4 factorial with the following factors and settings:

Factor	Settings	
	-1	$+1$
Head position during toss (Head)	Service line on opponent's court	In the air, at peak of balltoss
Back posture (Back)	Straight	Arched back
Right foot alignment (Foot)	Parallel to the baseline	60° angle from baseline
Breathing (Breath)	Inhale while striking	Exhale while striking

Results from the factorial are shown below:

Head	Back	Foot	Breath	Speed
-1	-1	-1	-1	26
-1	-1	-1	1	40
-1	-1	1	-1	29
-1	-1	1	1	39
-1	1	-1	-1	42
-1	1	-1	1	29
-1	1	1	-1	23
-1	1	1	1	30
1	-1	-1	-1	45
1	-1	-1	1	38
1	-1	1	-1	44
1	-1	1	1	51
1	1	-1	-1	55
1	1	-1	1	59
1	1	1	-1	66
1	1	1	1	50

(*a*) Estimate the effects and their interactions, and use a normal probability plot to make decisions regarding their significance.

(*b*) What recommendations would you make to this tennis player?

(*c*) Use ANOVA to estimate the effects of these factors by ignoring all but main effects and simple interactions. Are your conclusions the same as those in (a)?

(*d*) How would you have designed this experiment if the tennis player could only serve four serves at a time without fatigue?

8. You just overheard your new boss at HSC say, "This randomization stuff is baloney—it's sloppy experimenting and it wastes the operators' time." Prepare a 1-minute discourse to refute this statement and convince your boss of the value of randomization.

9. A greedy parent has decided to use her children to garner candy for her at Halloween, so she is interested in the factors which affect candy donations. She has a 13-year-old boy and a 7-year-old girl who will participate in the experiment. She selected a sample of 32 homes, and replicated a 2^2 experiment as follows:

Factors	Settings	
	-1	*+1*
Age of child (Age)	7	13
Costume	Scary	Cute

The experiment was conducted over a period of 4 hours on Halloween evening; treatment combinations were randomized. The data from that evening is shown below in terms of pieces of candy collected:

Age	Costume	Pieces of Candy							
-1	-1	22	28	24	27	27	18	17	19
-1	1	29	22	28	30	27	27	34	22
1	-1	17	18	19	23	18	24	18	14
1	1	7	14	9	12	18	8	9	7

(*a*) Use ANOVA to determine the effects of Age and Costume on the amount of candy donated.

(*b*) Are there any confounding factors which may have affected the outcome of this experiment? How would this affect the generalization of experiment results to other families?

★ 10. A bowler has decided to investigate the effects of two factors on his bowling scores. The factors and their settings are shown below:

	Settings	
Factor	*−1*	*+1*
Starting position	Third dot	Fourth dot
Starting foot	Left	Right

The experiment was conducted in a single evening, over the course of eight games. It was a hot night, so the bowler drank beer throughout the evening; on the advice of his teammates, "cumulative beers consumed" was measured and will be used as a covariate.

Data from the experiment is shown below:

Position	*Foot*	*Beer*	*Score*
−1	−1	7.25	87
−1	−1	3.25	119
−1	1	2.75	162
−1	1	10.75	98
1	−1	7.50	156
1	−1	2.50	196
1	1	2.50	244
1	1	1.50	252

(a) Do an analysis of covariance on this data, and test for the effect of position, foot, and beer consumption.

(b) Is the effect of beer consumption the same for both position settings? for both foot settings?

(c) What would your analysis have indicated if you had not accounted for beer consumption?

CHAPTER 5

SCREENING EXPERIMENTS

5.1 INTRODUCTION

Screening experiments are the among first efforts taken to understand a new process or machine. They allow the experimenter to simultaneously test many factors for their influence on important process responses, although they usually do not give very precise estimates for the magnitude of factor effects, and many screening designs do not even permit the estimation of interactions. The benefit of screening designs lies in their ability to expeditiously select the few truly important factors from the many possible choices; after these major influences have been discovered, more refined experiments can be conducted to estimate effects precisely and answer important questions regarding interactions.

Two different types of screening experiments will be described: fractional factorial experiments are explained in Sections 5.2 through 5.6, and Plackett-Burman designs are presented in Section 5.7. These two types of experiments should be sufficient for most screening needs.

Metal Etch

The low resistance of many metals makes them nearly ideal choices for long
distance interconnects on integrated circuits. Metal layers are usually among
the last to be fabricated on a wafer: most metals have much lower melting
points than semiconductor materials like silicon dioxide, so they cannot be
utilized before those materials are already deposited.

Metals lines are fabricated like most other layers:

- *Deposition:* One of the popular deposition techniques is *sputtering,* in which a
  wafer is placed under a metal "target" while that target is bombarded by a high-
  energy ionized gas, as shown in Figure 5.1. The gas molecules knock off bits of
  the target and "sputter" them onto the wafer. Metal films of high purity and uniformity can be obtained in this manner.
- *Lithographic patterning:* Metal lines are defined by patterned resist like any
  other layer.
- *Etching:* Metal not protected by resist is etched away, leaving only the desired
  metal lines.
- *Resist strip:* Any resist remaining on the wafer is removed with a suitable solvent.

The metal etch process itself can be very complicated. The examples in this
chapter relate to a process with the following etch procedure:

- Preclean the wafer. This step removes any residual metal oxides which would
  impede the following etch. Preclean solutions and dip times must be carefully
  controlled to avoid metal corrosion.
- Etch the wafer. The requires a multistep plasma etch. Times, temperatures, and
  the frequency at which gases are energized must be carefully controlled if an effective and uniform etch is to be obtained. Etch waste products are continuously
  monitored, and the end of the etch step is marked by a sudden drop (called an
  endpoint) in metal ions. A timed overetch is done to ensure that no unwanted
  residual metal remains.
- A postclean is done shortly after the etch to remove corrosive etch by-products
  that would otherwise corrode the metal lines.

In addition to the obvious need to produce a line of well-controlled width and
thickness, the profile of the line can also be very critical. This particular
process sometimes had a small metal "foot" at the bottom of the line. This
foot caused the capacitance between long, parallel lines to be large enough
to cause timing problems in the circuit.

5.2   FRACTIONAL FACTORIAL EXPERIMENTS

Fractional factorial experiments are carefully selected subsets of factorial experiments that can estimate the effects of factors for lower cost than full factorials. They vary in their ability to estimate interaction effects, but in general they are not designed to obtain estimates of higher-order interactions. Complex interactions are rare, and their magnitude is small relative to main effects when they do occur, so the inability of fractional factorials to estimate them is a small price to pay for the increased efficiency of fractional factorial designs.

Fractional factorial experiments are described with a concise notation similar to that of factorial experiments:

- A 2^{k-p} experiment has k factors.
- It is a $(\frac{1}{2})^p$ fraction of the full 2^k factorial.
- The number of runs in a single replicate of the experiment is 2^{k-p}.

So, for example, a 2^{11-5} fractional factorial has 11 factors, is a one 32nd fraction of the 2^{11} factorial, and requires 64 runs per replicate (2^6).

The 2^{3-1} is a simple fractional factorial that will be used to illustrate some general properties of these designs.

Recall the design of a 2^3 factorial experiment:

X1	X2	X3
−1	−1	−1
−1	−1	1
−1	1	−1
−1	1	1
1	−1	−1
1	−1	1
1	1	−1
1	1	1

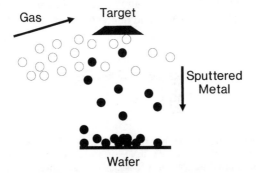

Fig. 5.1   Highly energized gas molecules (open circles) are directed at a metal target. They knock off metal molecules from the target (solid circles), which are then deposited on the wafer below.

This type of experiment varies each of three factors to a low (-1) and high (1) value, and all eight possible combinations of these levels are included in the experiment. The 2^3 factorial obtains effects estimates for the main effects, the simple interactions between those effects, and their three-way interaction:

$$X1 \quad X1*X2 \quad X1*X2*X3$$
$$X2 \quad X1*X3$$
$$X3 \quad X2*X3$$

A layout for a 2^{3-1} fractional factorial experiment with these same factors is shown below:

X1	X2	X3
-1	-1	1
-1	1	-1
1	-1	-1
1	1	1

Only four runs are required for a replicate of *this* experiment, but some of the effects that could be estimated with the 2^3 are no longer estimable:

X1	X2	X3	X1*X2
-1	-1	1	1
-1	1	-1	-1
1	-1	-1	-1
1	1	1	1

The X1*X2 interaction always takes the same value as X3, so their effects estimates are the same: both of these effects will appear to be significant if only one of them is actually significant. These effects are confounded, and in the terminology of screening designs, they are said to be *aliases* of one another.

The aliasing situation for this design is described visually in the *aliasing diagram* shown in Figure 5.2. Factors and interactions are listed at the left of the tri-

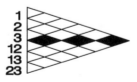

Fig. 5.2 The aliasing diagram for this 2^{3-1} resolution III fractional factorial design shows that each main effect (denoted 1, 2, and 3) is aliased with one simple interaction. Aliases for each effect are found by looking for shaded cells in the rows branching from that effect, and then following the intersecting row back to the left of the triangle to identify the alias. For example, the 1 effect is aliased with the 23 interaction because the intersection of their columns is shaded.

angle: 1 stands for factor 1, and 12 stands for the interaction of factor 1 and factor 2. Aliases for each effect are found by looking for shaded cells in the rows branching from that effect, and then following the intersecting row back to the left of the triangle to identify the alias.

■ *Example 5.1: Using the 2^{3-1} Aliasing Diagram*

To find the aliases of the 12 interaction, start in the 12 row and "look both ways": looking down the row reveals no aliases, but looking up the row reveals an alias with the 3 effect. To find the aliases of the 23 effect, look in the only direction possible (up) and find that the 1 effect is an alias.

   Aliasing diagrams are given for a variety of fractional factorial designs in Figures 5.2 through 5.20. Note that aliases with main effects are denoted with solid shading, and aliases involving only interactions are denoted with crosshatching. Aliases with higher-order effects do exist, but they are not noted in the aliasing diagrams.

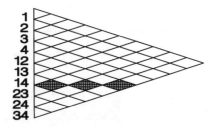

Fig. 5.3   The 2^{4-1} resolution IV design aliases some interactions with one another, as noted by crosshatching in intersecting columns for aliased effects.

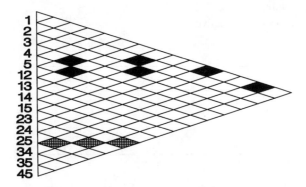

Fig. 5.4   The 2^{5-2} resolution III design aliases some main effects with interactions (solid shading) and some interactions with one another (crosshatched shading).

The particular choice of experimental runs used to construct a fractional factorial is very important, because some choices produce much less desirable aliasing. There are seventy ways to choose four runs from the eight runs in the 2^3 factorial, and 68 of those choices are unsatisfactory.

Consider the design below:

X1	*X2*	*X3*
−1	−1	−1
−1	−1	1
1	1	−1
1	1	1

X1 and X2 are aliased in this design—this is clearly unacceptable. Consider this alternative design:

X1	*X2*	*X3*
−1	1	−1
−1	1	1
1	−1	−1
1	−1	1

X1 and X2 are also aliased in this design because X2 is always the opposite of X1; if either factor has an effect, both factors will appear to have an effect. This is also an unacceptable design.

Consider one more alternative:

X1	*X2*	*X3*
−1	−1	−1
−1	1	−1
−1	1	1
1	−1	1

Main effects are also aliased in this design, although the pattern is less obvious. The estimate of the effect of X1, E1, is given by:

$$E2 = -\frac{3}{4} E1 - \frac{3}{4} E13 - E23$$

Another desirable property of the design has been sacrificed too: the design is no longer balanced because X1 takes its high value only one of four times.

Choosing the right combination of runs is not a trivial exercise, but tables or software can be used to select effective designs. The interested reader can consult Box et al. (1978) for details on the special algebra used to construct fractional factorials. Instructions for several practical designs are given in Table 5.1.

5.3 DESIGN OF FRACTIONAL FACTORIAL EXPERIMENTS

Screening designs are very similar to factorial experiments. Factors and levels for those factors must be chosen, and the considerations involved are essentially the same as for factorial experiments except that more factors can be included in the experiments. Centerpoints are used just as before. Clever use of blocking in fractional factorials can be used to great advantage, so Section 5.6 is dedicated to this topic.

Fractional factorials do introduce additional design complexity because of aliasing. This topic will be discussed in detail, but as a general rule, early experiments will have more factors and more aliasing than later experiments.

5.3.1 Number of Factors and Levels

Like most factorial experiments, factors in screening designs are typically varied to only two levels. If information on quadratic or higher-order effects is needed, response surface methods (Chapter 6) are a more cost-effective choice.

Screening designs permit simultaneous experimentation with a larger number of factors than factorials. The 2^{11-7} is a case in point—the corresponding factorial would have required 2048 runs per replicate; the fractional factorial uses 16 runs. There are practical limits on the number of factors, however:

- For experiments with 11 or more factors, Plackett-Burman designs are usually a more cost-effective choice than fractional factorials.
- Fractional factorials can be split across blocks (Section 5.6.2) just as factorials can, but it is often advantageous to run the entire suite of treatment combinations within a block. This places a natural upper limit on the number of factors allowed.
- Unknown confounding factors have a greater ability to mislead the experimenter in a highly fractionated factorial with many factors: the influences of multiple factors can be hidden or falsely inflated, and unless the experiment is replicated or many centerpoints are run, it will be impossible to detect such confounding.
- The more factors in an experiment, the more likely that at least some portion of it will not be correctly executed.

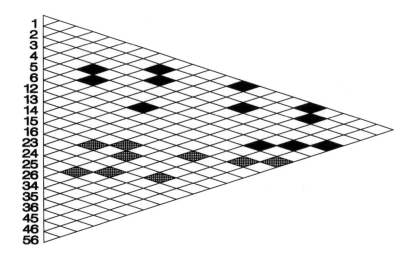

Fig. 5.5   Aliasing diagram for the 2^{6-3} resolution III design.

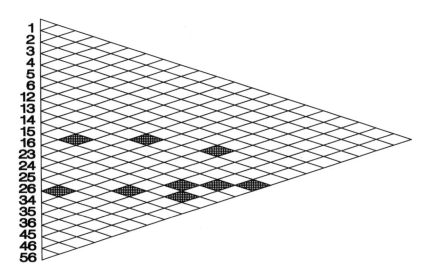

Fig. 5.6   Aliasing diagram for the 2^{6-2} resolution IV design.

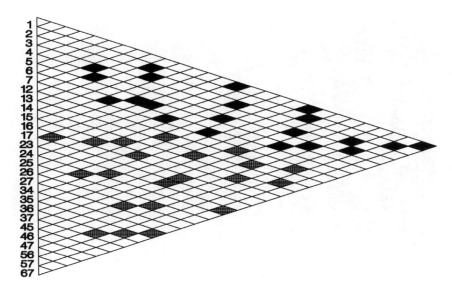

Fig. 5.7   Aliasing diagram for the 2^{7-4} resolution III design.

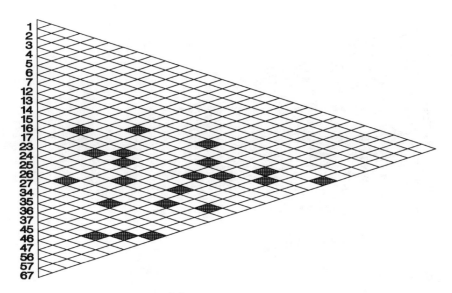

Fig. 5.8   Aliasing diagram for the 2^{7-3} resolution IV design.

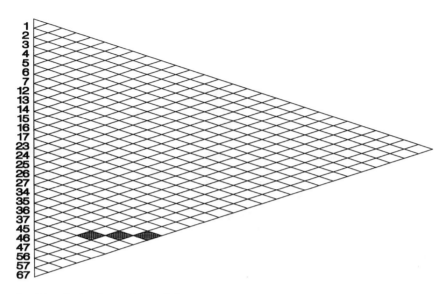

Fig. 5.9   Aliasing diagram for the 2^{7-2} resolution IV design.

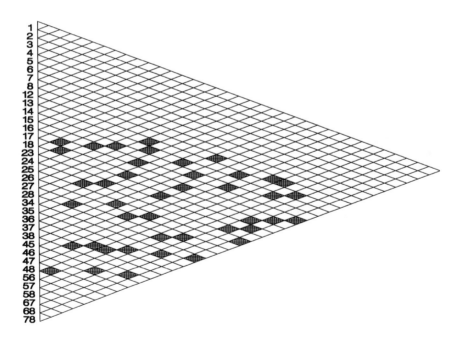

Fig. 5.10   Aliasing diagram for the 2^{8-4} resolution IV design.

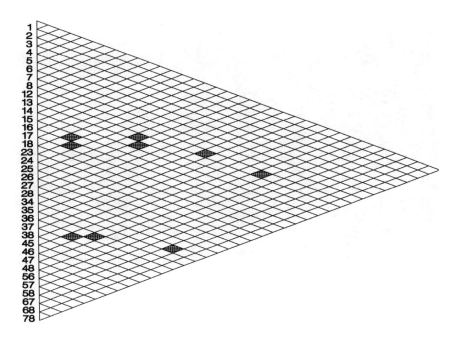

Fig. 5.11   Aliasing diagram for the 2^{8-3} resolution IV design.

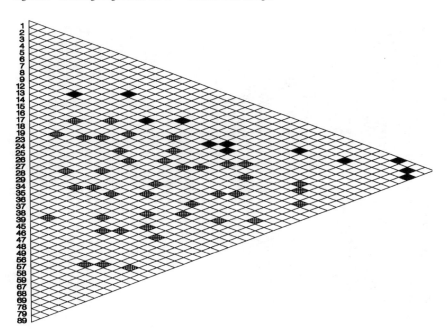

Fig. 5.12   Aliasing diagram for the 2^{9-5} resolution III design.

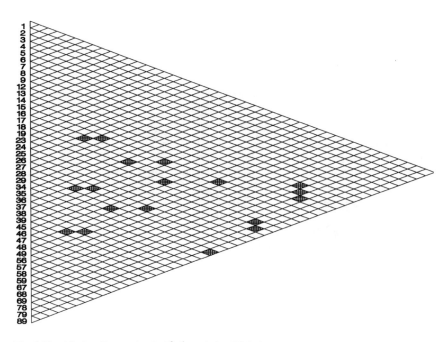

Fig. 5.13   Aliasing diagram for the 2^{9-4} resolution IV design.

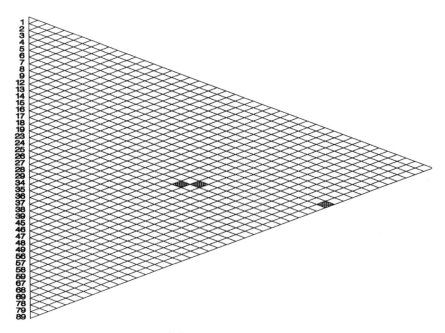

Fig. 5.14   Aliasing diagram for the 2^{9-3} resolution IV design.

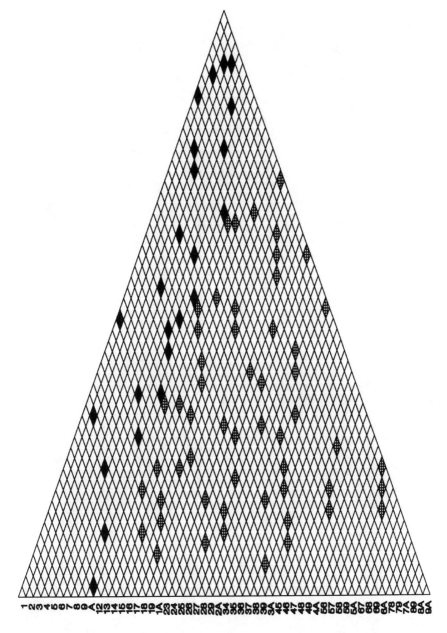

Fig. 5.15   Aliasing diagram for the 2^{10-6} resolution III design. The tenth factor is denoted by the letter "A."

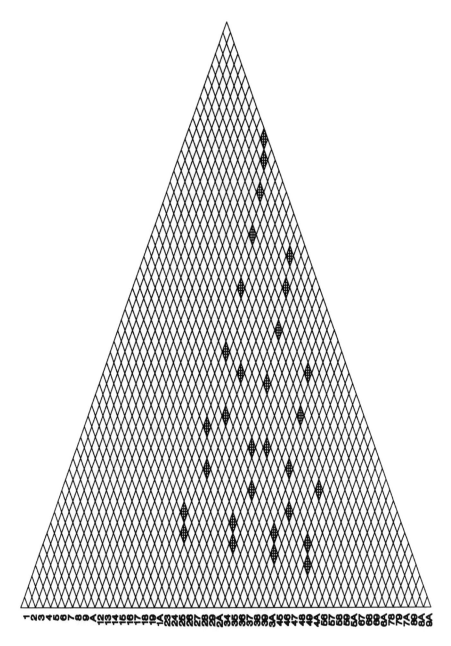

Fig. 5.16   Aliasing diagram for the 2^{10-5} resolution IV design.

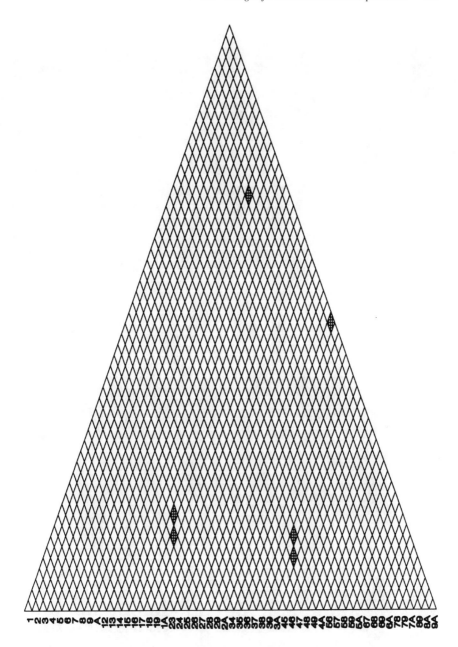

Fig. 5.17   Aliasing diagram for the 2^{10-4} resolution IV design.

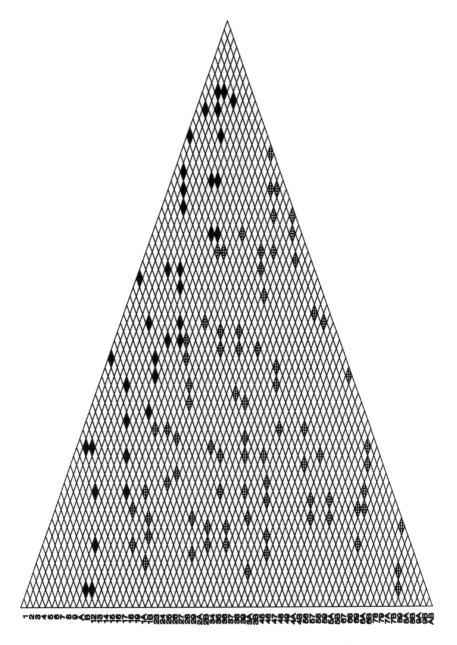

Fig. 5.18   Aliasing diagram for the 2^{11-7} resolution III design.

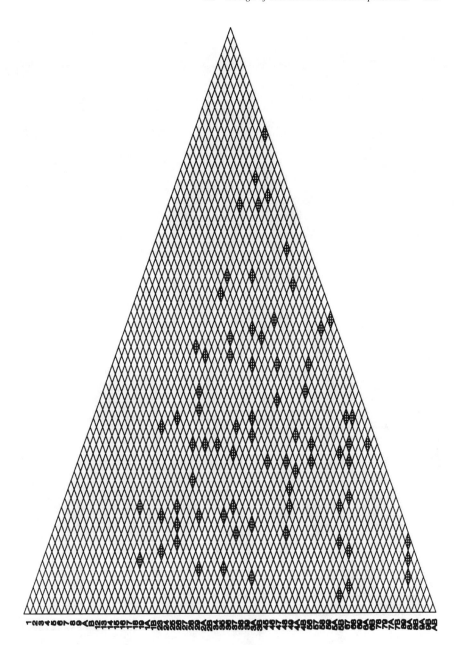

Fig. 5.19   Aliasing diagram for the 2^{11-6} resolution IV design.

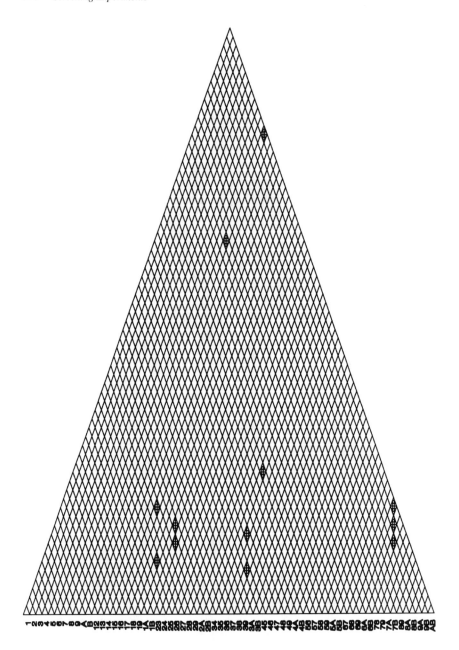

Fig. 5.20   Aliasing diagram for the 2^{11-5} resolution IV design.

TABLE 5.1   Fractional Factorial Designs*

2^{3-1} *Resolution: III*

1	2	3
−1	−1	1
−1	1	−1

2^{4-1} *Resolution: IV*

1	2	3	4
−1	−1	−1	−1
−1	−1	1	1
−1	1	−1	1
−1	1	1	−1
1	−1	−1	1
1	−1	1	−1
1	1	−1	−1
1	1	1	1

2^{5-2} *Resolution: III*

1	2	3	4	5
−1	−1	−1	1	1
−1	−1	1	1	−1
−1	1	−1	−1	1
−1	1	1	−1	−1
1	−1	−1	−1	−1
1	−1	1	−1	1
1	1	−1	1	−1
1	1	1	1	1

2^{5-1} *Resolution: V*

1	2	3	4	5
−1	−1	−1	−1	1
−1	−1	−1	1	−1
−1	−1	1	−1	−1
−1	−1	1	1	1
−1	1	−1	−1	−1
−1	1	−1	1	1
−1	1	1	−1	1
−1	1	1	1	−1
1	−1	−1	−1	−1
1	−1	−1	1	1
1	−1	1	−1	1
1	−1	1	1	−1
1	1	−1	−1	1
1	1	−1	1	−1
1	1	1	−1	−1
1	1	1	1	1

(Continued)

2^{6-3} *Resolution: III*

1	2	3	4	5	6
−1	−1	−1	1	1	1
−1	−1	1	1	−1	−1
−1	1	−1	−1	1	−1
−1	1	1	−1	−1	1
1	−1	−1	−1	−1	1
1	−1	1	−1	1	−1
1	1	−1	1	−1	−1
1	1	1	1	1	1

2^{6-2} *Resolution: IV*

1	2	3	4	5	6
−1	−1	−1	−1	−1	
−1	−1	−1	1	−1	1
−1	−1	1	−1	1	1
−1	−1	1	1	1	−1
−1	1	−1	−1	1	1
−1	1	−1	1	1	−1
−1	1	1	−1	−1	−1
−1	1	1	1	−1	1
1	−1	−1	−1	1	−1
1	−1	−1	1	1	1
1	−1	1	−1	−1	1
1	−1	1	1	−1	−1
1	1	−1	−1	−1	1
1	1	−1	1	−1	−1
1	1	1	−1	1	−1
1	1	1	1	1	1

2^{7-4} *Resolution: III*

1	2	3	4	5	6	7
−1	−1	−1	1	1	1	−1
−1	−1	1	1	−1	−1	1
−1	1	−1	−1	1	−1	1
−1	1	1	−1	−1	1	−1
1	−1	−1	−1	−1	1	1
1	−1	1	−1	1	−1	−1
1	1	−1	1	−1	−1	−1
1	1	1	1	1	1	1

(Continued)

2^{7-3} *Resolution IV*

1	2	3	4	5	6	7
−1	−1	−1	−1	−1	−1	−1
−1	−1	−1	1	−1	1	1
−1	−1	1	−1	1	1	1
−1	−1	1	1	1	−1	−1
−1	1	−1	−1	1	1	−1
−1	1	−1	1	1	−1	1
−1	1	1	−1	−1	−1	1
−1	1	1	1	−1	1	−1
1	−1	−1	−1	1	−1	1
1	−1	−1	1	1	1	−1
1	−1	1	−1	−1	1	−1
1	−1	1	1	−1	−1	1
1	1	−1	−1	−1	1	1
1	1	−1	1	−1	−1	−1
1	1	1	−1	1	−1	−1
1	1	1	1	1	1	1

2^{7-2} *Resolution: IV*

1	2	3	4	5	6	7
−1	−1	−1	−1	−1	1	1
−1	−1	−1	−1	1	1	−1
−1	−1	−1	1	−1	−1	−1
−1	−1	−1	1	1	−1	1
−1	−1	1	−1	−1	−1	1
−1	−1	1	−1	1	−1	−1
−1	−1	1	1	−1	1	−1
−1	−1	1	1	1	1	1
−1	1	−1	−1	−1	−1	−1
−1	1	−1	−1	1	−1	1
−1	1	−1	1	−1	1	1
−1	1	−1	1	1	1	−1
−1	1	1	−1	−1	1	−1
−1	1	1	−1	1	1	1
−1	1	1	1	−1	−1	1
−1	1	1	1	1	−1	−1
1	−1	−1	−1	−1	−1	−1
1	−1	−1	−1	1	−1	1
1	−1	−1	1	−1	1	1
1	−1	−1	1	1	1	−1
1	−1	1	−1	−1	1	−1
1	−1	1	−1	1	1	1
1	−1	1	1	−1	−1	1
1	−1	1	1	1	−1	−1
1	1	−1	−1	−1	1	1
1	1	−1	−1	1	1	−1

(Continued)

2^{7-2} Resolution: IV (Continued)

1	2	3	4	5	6	7
1	1	−1	1	−1	−1	−1
1	1	−1	1	1	−1	1
1	1	1	−1	−1	−1	1
1	1	1	−1	1	−1	−1
1	1	1	1	−1	1	−1
1	1	1	1	1	1	1

2^{8-4} Resolution: IV

1	2	3	4	5	6	7	8
−1	−1	−1	−1	−1	−1	−1	−1
−1	−1	−1	1	1	1	−1	1
−1	−1	1	−1	1	1	1	−1
−1	−1	1	1	−1	−1	1	1
−1	1	−1	−1	1	−1	1	1
−1	1	−1	1	−1	1	1	−1
−1	1	1	−1	−1	1	−1	1
−1	1	1	1	1	−1	−1	−1
1	−1	−1	−1	−1	1	1	1
1	−1	−1	1	1	−1	1	−1
1	−1	1	−1	1	−1	−1	1
1	−1	1	1	−1	1	−1	−1
1	1	−1	−1	1	1	−1	−1
1	1	−1	1	−1	−1	−1	1
1	1	1	−1	−1	−1	1	−1
1	1	1	1	1	1	1	1

2^{8-3} Resolution: IV

1	2	3	4	5	6	7	8
−1	−1	−1	−1	−1	−1	−1	1
−1	−1	−1	−1	1	−1	−1	−1
−1	−1	−1	1	−1	−1	1	−1
−1	−1	−1	1	1	−1	1	1
−1	−1	1	−1	−1	1	−1	−1
−1	−1	1	−1	1	1	−1	1
−1	−1	1	1	−1	1	1	1
−1	−1	1	1	1	1	1	−1
−1	1	−1	−1	−1	1	1	−1
−1	1	−1	−1	1	1	1	1
−1	1	−1	1	−1	1	−1	1
−1	1	−1	1	1	1	−1	−1
−1	1	1	−1	−1	−1	1	1
−1	1	1	−1	1	−1	1	−1
−1	1	1	1	−1	−1	−1	−1
−1	1	1	1	1	−1	−1	1
1	−1	−1	−1	−1	1	1	1
1	−1	−1	−1	1	1	1	−1
1	−1	−1	1	−1	1	−1	−1

(Continued)

2^{8-3} *Resolution: IV (Continued)*

1	2	3	4	5	6	7	8
1	-1	-1	1	1	1	-1	1
1	-1	1	-1	-1	-1	1	-1
1	-1	1	-1	1	-1	1	1
1	-1	1	1	-1	-1	-1	1
1	-1	1	1	1	-1	-1	-1
1	1	-1	-1	-1	-1	-1	-1
1	1	-1	-1	1	-1	-1	1
1	1	-1	1	-1	-1	1	1
1	1	-1	1	1	-1	1	-1
1	1	1	-1	-1	1	-1	1
1	1	1	-1	1	1	-1	-1
1	1	1	1	-1	1	1	-1
1	1	1	1	1	1	1	1

2^{8-2} *Resolution: V*

1	2	3	4	5	6	7	8
-1	-1	-1	-1	-1	-1	1	1
-1	-1	-1	-1	-1	1	1	-1
-1	-1	-1	-1	1	-1	1	-1
-1	-1	-1	-1	1	1	1	1
-1	-1	-1	1	-1	-1	-1	1
-1	-1	-1	1	-1	1	-1	-1
-1	-1	-1	1	1	-1	-1	-1
-1	-1	-1	1	1	1	-1	1
-1	-1	1	-1	-1	-1	-1	1
-1	-1	1	-1	-1	-1	-1	-1
-1	-1	1	-1	1	-1	-1	-1
-1	-1	1	-1	1	1	-1	1
-1	-1	1	1	-1	-1	1	1
-1	-1	1	1	-1	1	1	-1
-1	-1	1	1	1	-1	1	-1
-1	-1	1	1	1	1	1	1
-1	1	-1	-1	-1	-1	-1	-1
-1	1	-1	-1	-1	1	-1	1
-1	1	-1	-1	1	-1	-1	1
-1	1	-1	-1	1	1	-1	1
-1	1	-1	1	-1	-1	1	-1
-1	1	-1	1	-1	1	1	1
-1	1	-1	1	1	-1	1	1
-1	1	-1	1	1	1	1	-1
-1	1	1	-1	-1	-1	1	-1
-1	1	1	-1	-1	1	1	1
-1	1	1	-1	1	-1	1	1
-1	1	1	-1	1	1	1	-1
-1	1	1	1	-1	-1	-1	-1
-1	1	1	1	-1	1	-1	1
-1	1	1	1	1	-1	-1	1
-1	1	1	1	1	1	-1	-1

1	2	3	4	5	6	7	8
1	-1	-1	-1	-1	-1	-1	-1
1	-1	-1	-1	-1	1	-1	1
1	-1	-1	-1	1	-1	-1	1
1	-1	-1	-1	1	1	-1	-1
1	-1	-1	1	-1	-1	1	-1
1	-1	-1	1	-1	1	1	1
1	-1	-1	1	1	-1	1	1
1	-1	-1	1	1	1	1	-1
1	-1	1	-1	-1	-1	1	-1
1	-1	1	-1	-1	1	1	1
1	-1	1	-1	1	-1	1	1
1	-1	1	-1	1	1	1	-1
1	-1	1	1	-1	-1	-1	-1
1	-1	1	1	-1	1	-1	1
1	-1	1	1	1	-1	-1	1
1	-1	1	1	1	1	-1	-1
1	1	-1	-1	-1	-1	1	1
1	1	-1	-1	-1	1	1	-1
1	1	-1	-1	1	-1	1	-1
1	1	-1	-1	1	1	1	1
1	1	-1	1	-1	-1	-1	1
1	1	-1	1	-1	1	-1	-1
1	1	-1	1	1	-1	-1	-1
1	1	-1	1	1	1	-1	1
1	1	1	-1	-1	-1	-1	1
1	1	1	-1	-1	1	-1	-1
1	1	1	-1	1	-1	-1	-1
1	1	1	-1	1	1	-1	1
1	1	1	1	-1	-1	1	1
1	1	1	1	-1	1	1	-1
1	1	1	1	1	-1	1	-1
1	1	1	1	1	1	1	1

(Continued)

2^{9-5} *Resolution: III*

1	2	3	4	5	6	7	8	9
-1	-1	-1	-1	-1	-1	-1	-1	1
-1	-1	-1	1	-1	1	1	1	-1
-1	-1	1	-1	1	1	1	-1	-1
-1	-1	1	1	1	-1	-1	1	1
-1	1	-1	-1	1	1	-1	1	-1
-1	1	-1	1	1	-1	1	-1	1
-1	1	1	-1	-1	-1	1	1	1
-1	1	1	1	-1	1	-1	-1	-1
1	-1	-1	-1	1	-1	1	1	-1
1	-1	-1	1	1	1	-1	-1	1
1	-1	1	-1	-1	1	-1	1	1
1	-1	1	1	-1	-1	1	-1	-1
1	1	-1	-1	-1	1	1	-1	1
1	1	-1	1	-1	-1	-1	1	-1
1	1	1	-1	1	-1	-1	-1	-1
1	1	1	1	1	1	1	1	1

2^{9-4} *Resolution: IV*

1	2	3	4	5	6	7	8	9
-1	-1	-1	-1	-1	1	1	1	1
-1	-1	-1	-1	1	-1	-1	-1	-1
-1	-1	-1	1	-1	-1	-1	-1	1
-1	-1	-1	1	1	1	1	1	-1
-1	-1	1	-1	-1	-1	-1	1	-1
-1	-1	1	-1	1	1	1	-1	1
-1	-1	1	1	-1	1	1	-1	-1
-1	-1	1	1	1	-1	-1	1	1
-1	1	-1	-1	-1	-1	1	-1	-1
-1	1	-1	-1	1	1	-1	1	1
-1	1	-1	1	-1	1	-1	1	-1
-1	1	-1	1	1	-1	1	-1	1
-1	1	1	-1	-1	1	-1	-1	1
-1	1	1	-1	1	-1	1	1	-1
-1	1	1	1	-1	-1	1	1	1
-1	1	1	1	1	1	-1	-1	-1
1	-1	-1	-1	-1	1	-1	-1	-1
1	-1	-1	-1	1	-1	1	1	1
1	-1	-1	1	-1	-1	1	1	-1
1	-1	-1	1	1	1	-1	-1	1
1	-1	1	-1	-1	-1	1	-1	1
1	-1	1	-1	1	1	-1	1	-1
1	-1	1	1	-1	1	-1	1	1
1	-1	1	1	1	-1	1	-1	1
1	1	-1	-1	-1	-1	-1	1	1
1	1	-1	-1	1	1	1	-1	-1
1	1	-1	1	-1	1	1	-1	1

(Continued)

2^{9-4} Resolution: IV (Continued)

1	2	3	4	5	6	7	8	9
1	1	-1	1	1	-1	-1	1	-1
1	1	1	-1	-1	1	1	1	-1
1	1	1	-1	1	-1	-1	-1	1
1	1	1	1	-1	-1	-1	-1	-1
1	1	1	1	1	1	1	1	1

2^{9-3} Resolution: IV

1	2	3	4	5	6	7	8	9	1	2	3	4	5	6	7	8	9
-1	-1	-1	-1	-1	-1	1	1	1	1	-1	-1	-1	-1	-1	-1	-1	1
-1	-1	-1	-1	-1	1	1	-1	-1	1	-1	-1	-1	-1	1	-1	1	-1
-1	-1	-1	-1	1	-1	1	-1	-1	1	-1	-1	-1	1	-1	-1	1	-1
-1	-1	-1	-1	1	1	1	1	1	1	-1	-1	-1	1	1	-1	-1	1
-1	-1	-1	1	-1	-1	-1	1	-1	1	-1	-1	1	-1	-1	1	-1	-1
-1	-1	-1	1	-1	1	-1	-1	1	1	-1	-1	1	-1	1	1	1	1
-1	-1	-1	1	1	-1	-1	-1	1	1	-1	-1	1	1	-1	1	1	1
-1	-1	-1	1	1	1	-1	1	-1	1	-1	-1	1	1	1	1	-1	-1
-1	-1	1	-1	-1	-1	-1	-1	-1	1	-1	1	-1	-1	-1	1	1	-1
-1	-1	1	-1	-1	1	-1	1	1	1	-1	1	-1	-1	1	1	-1	1
-1	-1	1	-1	1	-1	-1	1	1	1	-1	1	-1	1	-1	1	-1	1
-1	-1	1	-1	1	1	-1	-1	-1	1	-1	1	-1	1	1	1	1	-1
-1	-1	1	1	-1	-1	1	-1	-1	1	-1	1	1	-1	-1	-1	1	1
-1	-1	1	1	-1	1	1	1	1	1	-1	1	1	-1	1	-1	-1	-1
-1	-1	1	1	1	-1	1	1	1	1	-1	1	1	1	-1	-1	-1	-1
-1	-1	1	1	1	1	1	-1	1	1	-1	1	1	1	1	-1	1	1
-1	1	-1	-1	-1	-1	-1	1	1	1	1	-1	-1	-1	-1	1	-1	1
-1	1	-1	-1	-1	1	-1	1	-1	1	1	-1	-1	-1	1	1	1	-1
-1	1	-1	-1	1	-1	-1	-1	-1	1	1	-1	-1	1	-1	1	1	-1
-1	1	-1	-1	1	1	-1	1	1	1	1	-1	-1	1	1	1	-1	1
-1	1	-1	1	-1	-1	1	1	-1	1	1	-1	1	-1	-1	-1	-1	-1
-1	1	-1	1	-1	1	1	-1	1	1	1	-1	1	-1	1	-1	1	1
-1	1	-1	1	1	-1	1	-1	1	1	1	-1	1	1	-1	-1	1	1
-1	1	-1	1	1	1	1	1	-1	1	1	-1	1	1	1	-1	-1	-1
-1	1	1	-1	-1	-1	1	-1	-1	1	1	1	-1	-1	-1	-1	1	-1
-1	1	1	-1	-1	1	1	1	1	1	1	1	-1	-1	1	-1	-1	1
-1	1	1	-1	1	-1	1	1	1	1	1	1	-1	1	-1	-1	-1	1
-1	1	1	-1	1	1	1	-1	-1	1	1	1	-1	1	1	-1	1	-1
-1	1	1	1	-1	-1	-1	-1	1	1	1	1	1	-1	-1	1	1	1
-1	1	1	1	-1	1	-1	1	-1	1	1	1	1	-1	1	1	-1	-1
-1	1	1	1	1	-1	-1	1	-1	1	1	1	1	1	-1	1	-1	-1
-1	1	1	1	1	1	-1	-1	1	1	1	1	1	1	1	1	1	1

(Continued)

2^{10-6} *Resolution: III*

1	2	3	4	5	6	7	8	9	A
-1	-1	-1	-1	-1	-1	-1	-1	1	1
-1	-1	-1	1	-1	1	1	1	-1	1
-1	-1	1	-1	1	1	1	-1	-1	1
-1	-1	1	1	1	-1	-1	1	1	1
-1	1	-1	-1	1	1	-1	1	-1	-1
-1	1	-1	1	1	-1	1	-1	1	-1
-1	1	1	-1	-1	-1	1	1	1	-1
-1	1	1	1	-1	1	-1	-1	-1	-1
1	-1	-1	-1	1	-1	1	1	-1	-1
1	-1	-1	1	1	1	-1	-1	1	-1
1	-1	1	-1	-1	1	-1	1	1	-1
1	-1	1	1	-1	-1	1	-1	-1	-1
1	1	-1	-1	-1	1	1	-1	1	1
1	1	-1	1	-1	-1	-1	1	-1	1
1	1	1	-1	1	-1	1	-1	-1	1
1	1	1	1	1	1	1	1	1	1

2^{10-5} *Resolution: IV*

1	2	3	4	5	6	7	8	9	A
-1	-1	-1	-1	-1	1	1	1	1	1
-1	-1	-1	-1	1	1	-1	-1	-1	
-1	-1	-1	1	-1	-1	1	-1	-1	-1
-1	-1	-1	1	1	-1	-1	1	1	1
-1	-1	1	-1	-1	-1	-1	1	-1	-1
-1	-1	1	-1	1	-1	1	-1	1	1
-1	-1	1	1	-1	1	-1	-1	1	1
-1	-1	1	1	1	1	1	1	-1	-1
-1	1	-1	-1	-1	-1	-1	-1	1	-1
-1	1	-1	-1	1	-1	1	1	-1	1
-1	1	-1	1	-1	1	-1	1	-1	1
-1	1	-1	1	1	1	1	-1	1	-1
-1	1	1	-1	-1	1	1	-1	-1	1
-1	1	1	-1	1	1	-1	1	1	-1
-1	1	1	1	-1	-1	1	1	1	-1
-1	1	1	1	1	-1	-1	-1	-1	1
1	-1	-1	-1	-1	-1	-1	-1	-1	1
1	-1	-1	-1	1	-1	1	1	1	-1
1	-1	-1	1	-1	1	-1	1	1	-1
1	-1	-1	1	1	1	1	-1	-1	1
1	-1	1	-1	-1	1	1	-1	1	-1
1	-1	1	-1	1	-1	1	1	-1	1
1	-1	1	1	-1	-1	1	1	-1	1
1	-1	1	1	1	-1	-1	-1	1	-1
1	1	-1	-1	-1	1	1	1	-1	-1
1	1	-1	-1	1	1	-1	-1	1	1
1	1	-1	1	-1	-1	1	-1	1	1

(Continued)

2^{10-5} Resolution: IV (Continued)

1	2	3	4	5	6	7	8	9	A
1	1	-1	1	1	-1	-1	1	-1	-1
1	1	1	-1	-1	-1	-1	1	1	1
1	1	1	-1	1	-1	1	-1	-1	-1
1	1	1	1	-1	1	-1	-1	-1	-1
1	1	1	1	1	1	1	1	1	1

2^{10-4} Resolution: IV

1	2	3	4	5	6	7	8	9	A	1	2	3	4	5	6	7	8	9	A
-1	-1	-1	-1	-1	-1	1	1	1	1	1	-1	-1	-1	-1	-1	1	-1	-1	-1
-1	-1	-1	-1	-1	1	-1	-1	1	1	1	-1	-1	-1	-1	1	-1	1	-1	-1
-1	-1	-1	-1	1	-1	1	1	-1	-1	1	-1	-1	-1	1	-1	1	-1	1	1
-1	-1	-1	-1	1	1	-1	-1	-1	-1	1	-1	-1	-1	1	1	-1	1	1	1
-1	-1	-1	1	-1	-1	-1	-1	-1	1	1	-1	-1	1	-1	-1	-1	1	1	-1
-1	-1	-1	1	-1	1	1	1	-1	1	1	-1	-1	1	-1	1	1	-1	1	-1
-1	-1	-1	1	1	-1	-1	-1	1	-1	1	-1	-1	1	1	-1	-1	1	-1	1
-1	-1	-1	1	1	1	1	1	1	-1	1	-1	-1	1	1	1	1	-1	-1	1
-1	-1	1	-1	-1	-1	-1	-1	1	-1	1	-1	1	-1	-1	-1	-1	1	-1	1
-1	-1	1	-1	-1	1	1	1	1	-1	1	-1	1	-1	-1	1	1	-1	-1	1
-1	-1	1	-1	1	-1	-1	-1	-1	1	1	-1	1	-1	1	-1	-1	1	1	-1
-1	-1	1	-1	1	1	1	1	-1	1	1	-1	1	-1	1	1	1	-1	1	-1
-1	-1	1	1	-1	-1	1	1	-1	-1	1	-1	1	1	-1	-1	1	-1	1	1
-1	-1	1	1	-1	1	-1	-1	-1	-1	1	-1	1	1	-1	1	-1	1	1	1
-1	-1	1	1	1	-1	1	1	1	1	1	-1	1	1	1	-1	1	-1	-1	-1
-1	-1	1	1	1	1	-1	-1	1	1	1	-1	1	1	1	1	-1	1	-1	-1
-1	1	-1	-1	-1	-1	-1	1	-1	-1	1	1	-1	-1	-1	-1	-1	-1	1	1
-1	1	-1	-1	-1	1	1	-1	-1	-1	1	1	-1	-1	-1	1	1	1	1	1
-1	1	-1	-1	1	-1	-1	1	1	1	1	1	-1	-1	1	-1	-1	-1	-1	-1
-1	1	-1	-1	1	1	1	-1	1	1	1	1	-1	-1	1	1	1	1	-1	-1
-1	1	-1	1	-1	-1	1	-1	1	-1	1	1	-1	1	-1	-1	1	1	-1	1
-1	1	-1	1	-1	1	-1	1	1	-1	1	1	-1	1	-1	1	-1	-1	-1	1
-1	1	-1	1	1	-1	1	-1	-1	1	1	1	-1	1	1	-1	1	1	1	-1
-1	1	-1	1	1	1	-1	1	-1	1	1	1	-1	1	1	1	-1	-1	1	-1
-1	1	1	-1	-1	-1	1	-1	-1	1	1	1	1	-1	-1	-1	1	1	1	-1
-1	1	1	-1	-1	1	-1	1	-1	1	1	1	1	-1	-1	1	-1	-1	1	-1
-1	1	1	-1	1	-1	1	-1	1	-1	1	1	1	-1	1	-1	1	1	-1	1
-1	1	1	-1	1	1	-1	1	1	-1	1	1	1	-1	1	1	-1	-1	-1	1
-1	1	1	1	-1	-1	-1	1	1	1	1	1	1	1	-1	-1	-1	-1	-1	-1
-1	1	1	1	-1	1	1	-1	1	1	1	1	1	1	-1	1	1	1	-1	-1
-1	1	1	1	1	-1	-1	1	-1	-1	1	1	1	1	1	-1	-1	-1	1	1
-1	1	1	1	1	1	1	-1	-1	-1	1	1	1	1	1	1	1	1	1	1

(Continued)

$$2^{11-7} \text{ Resolution: III}$$

1	2	3	4	5	6	7	8	9	A	B
-1	-1	-1	-1	-1	-1	-1	-1	1	1	1
-1	-1	-1	1	-1	1	1	1	-1	1	1
-1	-1	1	-1	1	1	1	-1	-1	1	-1
-1	-1	1	1	1	-1	-1	1	1	1	-1
-1	1	-1	-1	1	1	-1	1	-1	-1	1
-1	1	-1	1	1	-1	1	-1	1	-1	1
-1	1	1	-1	-1	-1	1	1	1	-1	-1
-1	1	1	1	-1	1	-1	-1	-1	-1	-1
1	-1	-1	-1	1	-1	1	1	-1	-1	-1
1	-1	-1	1	1	1	-1	-1	1	-1	-1
1	-1	1	-1	-1	1	-1	1	1	-1	1
1	-1	1	1	-1	-1	1	-1	-1	-1	1
1	1	-1	-1	-1	1	1	-1	1	1	-1
1	1	-1	1	-1	-1	-1	1	-1	1	-1
1	1	1	-1	1	-1	-1	-1	-1	1	1
1	1	1	1	1	1	1	1	1	1	1

$$2^{11-6} \text{ Resolution: IV}$$

1	2	3	4	5	6	7	8	9	A	B
-1	-1	-1	-1	-1	-1	-1	-1	-1	-1	-1
-1	-1	-1	-1	1	-1	-1	1	-1	1	1
-1	-1	-1	1	-1	-1	1	1	1	1	1
-1	-1	-1	1	1	-1	1	-1	1	-1	-1
-1	-1	1	-1	-1	1	1	1	1	-1	-1
-1	-1	1	-1	1	1	1	-1	1	1	1
-1	-1	1	1	-1	1	-1	-1	-1	1	1
-1	-1	1	1	1	1	-1	1	-1	-1	-1
-1	1	-1	-1	-1	1	1	-1	-1	-1	1
-1	1	-1	-1	1	1	1	1	-1	1	-1
-1	1	-1	1	-1	1	-1	1	1	1	-1
-1	1	-1	1	1	1	-1	-1	1	-1	1
-1	1	1	-1	-1	-1	-1	1	1	-1	1
-1	1	1	-1	1	-1	-1	-1	1	1	-1
-1	1	1	1	-1	-1	1	-1	-1	1	-1
-1	1	1	1	1	-1	1	1	-1	-1	1
1	-1	-1	-1	-1	1	-1	-1	1	1	-1
1	-1	-1	-1	1	1	-1	1	1	-1	1
1	-1	-1	1	-1	1	1	1	-1	-1	1
1	-1	-1	1	1	1	1	-1	-1	1	-1
1	-1	1	-1	-1	-1	1	1	-1	1	-1
1	-1	1	-1	1	-1	1	-1	-1	-1	1
1	-1	1	1	-1	-1	-1	-1	1	-1	1
1	-1	1	1	1	-1	-1	1	1	1	-1
1	1	-1	-1	-1	-1	1	-1	1	1	1

(Continued)

2^{11-6} *Resolution: IV (Continued)*

1	2	3	4	5	6	7	8	9	A	B
1	1	-1	-1	1	-1	1	1	1	-1	-1
1	1	-1	1	-1	-1	-1	1	-1	-1	-1
1	1	-1	1	1	-1	-1	-1	-1	1	1
1	1	1	-1	-1	1	-1	1	-1	1	1
1	1	1	-1	1	1	-1	-1	-1	-1	-1
1	1	1	1	-1	1	1	-1	1	-1	-1
1	1	1	1	1	1	1	1	1	1	1

2^{11-5} *Resolution: IV*

1	2	3	4	5	6	7	8	9	A	B
-1	-1	-1	-1	-1	-1	-1	1	-1	1	1
-1	-1	-1	-1	-1	1	-1	1	1	-1	1
-1	-1	-1	-1	1	-1	1	1	-1	-1	-1
-1	-1	-1	-1	1	1	1	1	1	1	1
-1	-1	-1	1	-1	-1	1	-1	-1	-1	-1
-1	-1	-1	1	-1	1	1	-1	1	1	1
-1	-1	-1	1	1	-1	-1	-1	1	1	1
-1	-1	-1	1	1	1	-1	-1	1	-1	-1
-1	-1	1	-1	-1	-1	1	-1	-1	1	1
-1	-1	1	-1	-1	1	1	-1	1	-1	-1
-1	-1	1	-1	1	-1	-1	-1	-1	-1	-1
-1	-1	1	-1	1	1	-1	-1	-1	1	1
-1	-1	1	1	-1	-1	-1	1	-1	-1	-1
-1	-1	1	1	-1	1	-1	1	1	1	1
-1	-1	1	1	1	-1	1	1	-1	1	1
-1	-1	1	1	1	1	1	1	1	-1	-1
-1	1	-1	-1	-1	-1	-1	-1	-1	-1	1
-1	1	-1	-1	-1	1	-1	-1	1	1	-1
-1	1	-1	-1	1	-1	1	-1	-1	1	-1
-1	1	-1	-1	1	1	1	-1	1	-1	1
-1	1	-1	1	-1	-1	1	1	-1	1	-1
-1	1	-1	1	-1	1	1	1	1	-1	1
-1	1	-1	1	1	-1	-1	1	-1	-1	1
-1	1	-1	1	1	1	-1	1	1	1	-1
-1	1	1	-1	-1	-1	1	1	-1	-1	1
-1	1	1	-1	-1	1	1	1	1	1	-1
-1	1	1	-1	1	-1	-1	1	-1	1	-1
-1	1	1	-1	1	1	-1	1	1	-1	1
-1	1	1	1	-1	-1	-1	-1	-1	-1	-1
-1	1	1	1	-1	1	-1	-1	1	-1	1
-1	1	1	1	1	-1	1	-1	-1	-1	1
-1	1	1	1	1	1	1	-1	1	1	-1

1	2	3	4	5	6	7	8	9	A	B
1	-1	-1	-1	-1	-1	-1	-1	1	1	-1
1	-1	-1	-1	-1	1	-1	-1	-1	-1	1
1	-1	-1	-1	1	-1	1	-1	1	-1	1
1	-1	-1	-1	1	1	1	-1	-1	1	-1
1	-1	-1	1	-1	-1	1	1	1	-1	1
1	-1	-1	1	-1	1	1	1	-1	1	-1
1	-1	-1	1	1	-1	-1	1	1	1	-1
1	-1	-1	1	1	1	-1	1	-1	-1	1
1	-1	1	-1	-1	-1	1	1	1	1	-1
1	-1	1	-1	-1	1	1	1	-1	-1	1
1	-1	1	-1	1	-1	-1	1	1	-1	1
1	-1	1	-1	1	1	-1	1	-1	1	-1
1	-1	1	1	-1	-1	-1	-1	1	-1	1
1	-1	1	1	-1	1	-1	-1	-1	1	-1
1	-1	1	1	1	-1	1	-1	1	1	-1
1	-1	1	1	1	1	1	-1	-1	-1	1
1	1	-1	-1	-1	-1	1	-1	-1	-1	-1
1	1	-1	-1	-1	1	1	1	1	1	1
1	1	-1	-1	1	-1	1	1	-1	1	1
1	1	-1	-1	1	1	1	1	1	-1	-1
1	1	-1	1	-1	-1	1	-1	-1	1	1
1	1	-1	1	-1	1	1	-1	1	-1	-1
1	1	-1	1	1	-1	-1	-1	-1	-1	-1
1	1	-1	1	1	1	-1	-1	1	1	1
1	1	1	-1	-1	-1	1	-1	-1	-1	-1
1	1	1	-1	-1	1	1	-1	1	1	1
1	1	1	-1	1	-1	-1	-1	-1	1	1
1	1	1	-1	1	1	-1	-1	1	-1	-1
1	1	1	1	-1	-1	-1	1	-1	1	1
1	1	1	1	-1	1	-1	1	1	-1	-1
1	1	1	1	1	-1	1	1	-1	-1	-1
1	1	1	1	1	1	1	1	1	1	1

*Layouts are given for a variety of 2^{p-k} resolution III through resolution V fractional factorial experiments. In each layout, the factors are listed at the top of the table, and the values (-1 or 1) to be taken by the factors are found below. Each row of the table corresponds to one experimental treatment combination. For example, the 2^{3-1} involves three factors, and requires four treatment combinations for each replicate. The first treatment combination is realized with the first and second factors take their low values, and the third factor takes its high value. No indication of aliasing is given in these layouts; refer to the appropriate aliasing diagram for this information.

5.3.2 Resolution

Fractional factorials differ in the way they alias effects with one another; most practical designs may be put into one of three classes depending on their resolution.

- Resolution III designs alias simple interactions with at least some main effects (Figure 5.21). The 2^{3-1} is a resolution III design.
- Resolution IV designs alias at least some simple interactions with other simple interactions (Figure 5.22). The 2^{4-1} is a resolution IV design, as seen from its aliasing diagram in Figure 5.3.
- Resolution V designs alias neither simple interactions nor main effects with each other (Figure 5.23). The 2^{5-1} is a resolution V design (so no aliasing diagram is given for it).

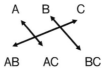

Fig. 5.21   In resolution III designs, main effects (A, B, and C here) are aliased with simple interactions (AB, AC, and BC). Simple interactions are also aliased with one another.

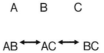

Fig. 5.22   Resolution IV designs alias at least some simple interactions with other simple interactions, but main effects are not aliased with simple interactions, nor with one another.

Fig. 5.23   Resolution V designs alias neither simple interactions nor main effects with each other. Aliasing of both main effects and simple interactions with higher-order interactions usually exists.

A summary of resolution for some common designs is shown in Figure 5.24. In all of these designs, higher-order interactions may be aliased with main effects or simple interactions.

Factors	Runs				
	4	8	16	32	64
3	3-1 / III	Unnecessary Expense			
4		4-1 / IV			
5	Unacceptable Aliasing	5-2 / III	5-1 / V		
6		6-3 / III	6-2 / IV		
7		7-4 / III	7-3 / IV	7-2 / IV	
8			8-4 / IV	8-3 / IV	8-2 / V
9			9-5 / III	9-4 / IV	9-3 / IV
10			10-6 / III	10-5 / IV	10-4 / IV
11			11-7 / III	11-6 / IV	11-5 / IV

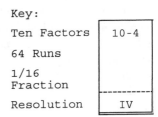

Key:

Ten Factors

64 Runs

1/16 Fraction

Resolution

10-4
IV

Fig. 5.24 This summary of fractional factorial designs helps choose experiments which have sufficient resolution, but minimum cost. Designs with resolution greater than V are in the "Unnecessary Expense" part of the table; those with less than resolution III are in the "Unacceptable Aliasing" part of the table. Each cell of the table concisely describes important aspects of the design. The top of the cell name the design ($10 - 4$ refers to a 2^{10-4} fractional factorial), and the bottom of the cell gives the resolution.

For any given number of factors, there will be a variety of resolutions from which to choose. The choice of resolution depends on both the cost allowed for the experiment, and on the type of information needed. For example, if five factors must be investigated, candidate designs would be the 2^{5-1} (resolution V), or the 2^{5-2} (resolution III). The 2^{5-3} design is economical (only four runs) but is generally unacceptable because it aliases main effects with one another. Resolution II and other designs that alias main effects with one another are known as *supersaturated* designs. They do find application in screening situations where there is reason to believe that most factors are not significant—the evaluation of novel therapeutic drugs is one example of such an applications. For further information see Booth and Cox (1962) or Watson (1961).

The concept of resolution simplifies the choice of design: if it is necessary to estimate main effects independently of interactions, a design of at least resolution IV (the 2^{5-1}) must be selected. If there is interest only in the main effects, the resolution III design will suffice.

Process characterization is best done in a sequence of experiments, with each one building on the knowledge obtained from previous experiments. When screening experiments are first initiated, resolution III designs are generally acceptable, since there are many possible factors to investigate. As ignorance declines as a result of experimentation, so does the number of important factors. Those factors which are found to be significant play a part in higher resolution experiments later in the sequence of experiments. A high-resolution experiment early in the experimentation sequence would be wasted, because it would estimate the interaction effects of many nonsignificant factors.

There are levels of resolution greater than V, but they are usually not cost-effective screening experiments because resolution V experiments with the same number of runs can obtain information on even more factors.

The 2^{7-1} is a resolution VII design requiring 64 runs. It does not alias main effects or interactions with anything less than a fifth-order interaction (12 is aliased with 34567), and in fact, even third-order interactions can be estimated with the 2^{7-1}. The 2^{8-2} design has resolution V, investigates an additional factor, and does so in the same number of runs. If a resolution IV design were acceptable in this situation, 11 factors could be included in a 2^{11-5} experiment of the same size. If a resolution III design were acceptable, 23 factors could be varied in a 24-run Plackett-Burman design.

These resolution-efficiency tradeoffs are summarized in the table below:

Design	Factors Tested	Resolution
2^{7-1}	7	VII
2^{8-2}	8	V
2^{11-5}	11	VI
PB-24	23	III

Understanding and controlling aliasing is the most exacting task in designing fractional factorial experiments. Careful use of aliasing diagrams can help to squeeze a little extra information out of fractional factorial experiments. For example, the 2^{6-2} design aliases the 13 interaction only with the 25 interaction. If factors can be chosen in such a way that the 25 interaction is either physically impossible or very unlikely, then a good estimate of the 13 interaction effect may be obtained.

5.3.3 Replication

Screening experiments are not designed to meet the same stringent beta risk goals that are set for factorial designs, so fractional factorial experiments are often not replicated.

Some replication is still a good idea if material constraints allow it:

- Replication allows the experimenter to assess the variance of a response at each combination of factor settings, thus helping to choose more robust process settings.
- It provides a convenient measure of within-group variance, thus facilitating statistical tests.
- It helps in checking model assumptions like homogeneity of variance and normality of residuals.
- It helps ensure that every desired treatment combination will be realized, even if some runs are lost during the course of experimentation.
- If a replicates can be run within a natural block, then replication is advisable to guard against making erroneous conclusions on the basis of one random—and perhaps very atypical—block.

There may also be good reasons *not* to replicate fractional factorial experiments. If the time from the start of an experiment to analysis of results is short, fractional factorials can be used very effectively in a sequential manner. Section 5.6.3 has details on this strategy.

Centerpoints are a very useful addition to fractional factorial experiments. Because these experiments are even more complicated than factorials, and usually involve more factors, an experimental control (centerpoint) is especially vital. Centerpoints also allow for some model assumptions to be checked, and this is impossible in completely unreplicated experiments. Centerpoints also provide an estimate of within-group variance which is used to perform objective tests of the statistical significance of effects.

5.4 EXECUTION

Since more factors are likely to be involved in fractional factorials than in factorials, there will be even more resistance to randomization than before. There are certainly situations in which some reasonable compromises will have to be made; however, screening experiments are even more susceptible to confounding than factorial experiments, so run order must be randomized as much as possible. This additional risk is due to the fact that only a relatively few experimental measurements are being used to estimate many effects, so one confounding factor can have a profound effect on conclusions about many factors. This book and most others show experiments in some "standard" order purely as a matter of convenience—it does not mean that randomization is unnecessary.

Centerpoints must be randomized along with the rest of the experiment; they provide the only means to check for confounding in unreplicated experiments.

5.5   ANALYSIS AND INTERPRETATION

Analysis of fractional factorials is almost exactly like that of factorials, with two exceptions:

- Depending on the degree of aliasing, the effect of some interaction terms cannot be estimated.
- The interpretation of statistical tests must be undertaken with full knowledge of the effects of aliasing.

The discussion of fractional factorials run within or across blocks is delayed to Section 5.6.

5.5.1   Experiments Including Replicates

The estimate of an effect is the same as it was for factorials: the average of responses at the high setting of a factor (or interaction) minus the average of the responses at the low setting of the factor.

■ *Example 5.2: Estimating Effects in a 2^{3-1} Experiment*

A 2^{3-1} fractional factorial design was part of a series of experiments on the metal preclean process.

The factors and levels chosen for this experiment were:

- Preclean solution (PRESOL): water (−) or methanol (+)
- Preclean dip time (PREDT): 90 and 120 seconds
- Preclean contains choline (+) or not (−) (PRECHO)

The response of interest was capacitance (CAP) measured between metal lines in a test structure.

The layout for one replicate of the experiment is shown below; simple interactions are computed to show the aliasing structure of the design.

PRESOL	PREDT	PRECHO	PRESOL*PREDT	PRESOL*PRECHO	PREDT*PTRCHO
−1	−1	1	1	−1	−1
−1	1	−1	−1	1	−1
1	−1	−1	−1	−1	1
1	1	1	1	1	1

The experiment was replicated three times, with the results shown below. The experiment was randomized, and the true run order is shown in the RUNORD column:

PRESOL	PREDT	PRECHO	RUNORD	CAP
−1	−1	1	8	166
−1	−1	1	9	160
−1	−1	1	11	177
−1	1	−1	1	176
−1	1	−1	3	178
−1	1	−1	7	210
1	−1	−1	4	162
1	−1	−1	5	184
1	−1	−1	6	182
1	1	1	2	137
1	1	1	10	156
1	1	1	12	164

Factor effects estimates can be computed directly from a table of means by factor:

Level	PRESOL Mean	PREDT Mean	PRECHO Mean
−1	177.833	171.833	182.000
1	164.167	170.167	160.000
Effect	−13.666	−1.666	−22.000

Estimates of the interaction effects are not available, because each interaction is aliased with a factor.

Statistical computing packages can easily perform statistical tests and compute estimated effects for fractional factorial experiments. A SAS analysis of the data above is shown in the example below.

■ *Example 5.3: SAS Analysis of a 2^{3-1} Experiment*

An analysis of variance of the data in the previous experiment was performed with SAS; edited output is shown below:

Dependent Variable: CAP				
Source	DF	Sum of Squares	Mean Square	Pr > F
Model	3	2020.666667	673.555556	0.0712
Error	8	1557.333333	194.666667	
Corrected total	11	3578.000000		
Source	DF	Type III SS	Mean Square	Pr > F
PRESOL	1	560.333333	560.333333	0.1282
PREDT	1	8.333333	8.333333	0.8413
PRECHO	1	1452.000000	1452.000000	0.0258

Apparently, the addition of choline to the preclean (PRECHO) has a significant effect on capacitance.

Tests for the interactions were not requested because all simple interactions are aliased with main effects.

In this example, tests for simple interactions were not requested because they are aliased with main effects.

Experiments with higher resolution can estimate simple interactions as is shown in the analysis of a 2^{5-1} design in the example below.

■ *Example 5.4: Analysis of a 2^{5-1} Experiment*

In this experiment the effects of five factors *and their simple interactions* were estimated:

- Preclean solution (PRESOL): water (–) or methanol (+)
- Etch gas 2 (EGAS): CCl_4 (–) or $CHCl_3$ (+)
- Etch temperature (ETEMP): 225 or 255°C
- Etch endpoint method (EENDP): Al ion (–) or Zr ion (+)
- Postclean method (POSTME): multiple quench-rinse (–) or washer-drier (+)

The experiment was not replicated, but some centerpoints were included to provide an estimate of within-group variance; all runs—including centerpoints—were randomized. The results of the experiment are shown in standardized order below.

PRESOL	EGAS	ETEMP	EENDP	POSTME	CAP
−1	−1	−1	−1	1	192
−1	−1	−1	1	−1	198
−1	−1	1	−1	−1	150
−1	−1	1	1	1	261
−1	1	−1	−1	−1	164
−1	1	−1	1	1	212
−1	1	1	−1	1	201
−1	1	1	1	−1	235
0	0	0	0	0	203
0	0	0	0	0	207
0	0	0	0	0	195
0	0	0	0	0	192
1	−1	−1	−1	−1	189
1	−1	−1	1	1	202
1	−1	1	−1	1	170
1	−1	1	1	−1	203
1	1	−1	−1	1	202
1	1	−1	1	−1	190
1	1	1	−1	−1	157
1	1	1	1	1	237

ANOVA was performed with SAS to test the significance of all main effects and simple interactions. The only measure of experimental error is furnished by the four centerpoints, so the tests could not detect small differences.

General Linear Models Procedure—Dependent Variable: CAP

Source	DF	Sum of Squares	Pr > F
Model	16	13233.2500000	0.0191
Error	3	144.7500000	
Corrected Total	19	13378.0000000	

Source	DF	Type III SS	Pr > F
PRESOL	1	248.06250000	0.1082
EGAS	1	68.06250000	0.3204
ETEMP	1	264.06250000	0.1013
EENDP	1	6123.06250000	0.0015
POSTME	1	2280.06250000	0.0063
PRESOL*EGAS	1	7.56250000	0.7187
PRESOL*ETEMP	1	588.06250000	0.0397
PRESOL*EENDP	1	451.56250000	0.0550
PRESOL*POSTME	1	138.06250000	0.1893
EGAS*ETEMP	1	217.56250000	0.1238
EGAS*EENDP	1	10.56250000	0.6718
EGAS*POSTME	1	27.56250000	0.5047
ETEMP*EENDP	1	2575.56250000	0.0053
ETEMP*POSTME	1	203.06250000	0.1326
EENDP*POSTME	1	22.56250000	0.5432

The significant factors and interactions in this case appear to be EENDP, POSTME, PRESOL*ETEMP, and ETEMP*EENDP.

Assumption checking proceeds exactly as before, and in replicated experiments residuals are plentiful, so serious problems are easy to recognize.

■ *Example 5.5: Assumption checking of a 2^{3-1} Experiment*

The 2^{3-1} experiment was replicated, so ample information is available for assumption checking. The residual analysis for that experiment includes a histogram of residuals (Figure 5.25), which seems to confirm that residuals are normally distributed.

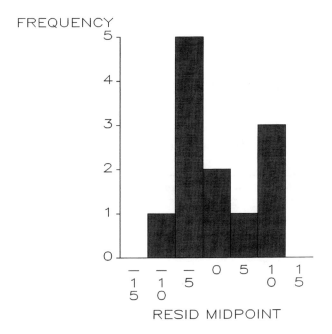

Fig. 5.25   This histogram of residuals from a 2^{3-1} fractional factorial experiment seems to confirm that residuals are normally distributed.

Boxplots of residuals by the first factor (Figure 5.26) are used to check for homogeneity of variance, and a plot of residuals in time order (Figure 5.27) does not show any persistent trends or patterns.

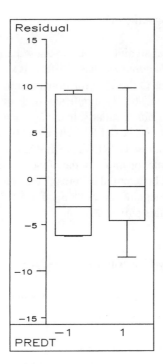

Fig. 5.26   These boxplots of residuals by one experimental factor seem to verify homogeneity of variance.

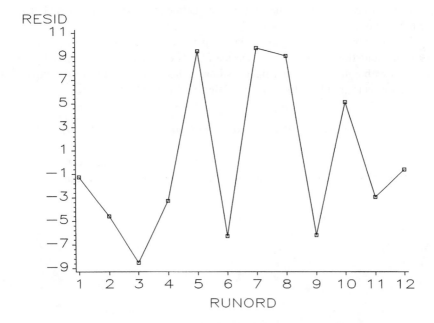

Fig. 5.27   This plot of residuals in time order does not show any trends or patterns, so the assumption of residual independence seems valid.

5.5.2 Interpreting Aliased Effects

The apparent statistical significance of an aliased effect is actually an indication that *one or more of the group* of aliased effects is significant. For example, the 12, 35, and 67 interactions of the 2^{7-3} fractional factorial are aliased, so if the 12 effect were found to be statistically significant, this could be due to any or all of the 12, 35, and 67 interactions. Similarly, in the 2^{3-1} experiment mentioned in examples, the apparent significance of PRECHO could be due to that factor, or to the interaction to which it is aliased (PREDT*PRESOL).

Unless the physical situation makes some of the aliased effects impossible, further experimentation will be required to "break" the alias and determine which effect is responsible for the significant result. Section 5.6.3 shows how to design additional experiments for this purpose.

5.5.3 Completely Unreplicated Designs

Completely unreplicated fractional factorial designs can be analyzed just as unreplicated factorial experiments were:

- By visual evaluation of a probability plot of effects estimates.
- By collapsing (pretending some factors were not varied) so artificial replication occurs.
- By assuming that particular interactions or other effects are unimportant.

These last two methods can cause an immoderate increase in experimental risks, so they should be used with caution.

The probability plot analysis method depends on the assumption that, if no effects are statistically significant, then the effects estimates will be normally distributed with a mean of zero. Effects estimates that seem far from zero compared to other effects are likely to signal truly significant effects.

■ *Example 5.6: Probability Plot Analysis of an Unreplicated 2^{5-1} Experiment*

Probability plot analysis was used to interpret the 2^{5-1} experiment with the centerpoints removed. Effects estimates are shown below in ascending order of effect:

Effect	Estimate
13	−12.125
14	−10.625
1	−7.875
15	−5.875
45	−2.375
24	−1.625
12	1.375
25	2.625
2	4.125
35	7.125
23	7.375
3	8.125
5	23.875
34	25.375
4	39.125

A probability plot of these estimates is shown in Figure 5.28. The three largest effects (for 4 = EENDP, 5 = POSTME, and the 34 = ETEMP*EENDP interaction) vary so much from the line determined by rest of the estimates that they are surely significant. The PRESOL*ETEMP interaction may also be significant because it falls below the fifth percentile of the distribution.

Fractional factorials can be converted to factorials by pretending that some of the factors are unimportant. For example, a 2^{3-1} can be collapsed over the third factor to produce a 2^2 in the first two factors:

2^{3-1}			2^2	
1	2	3	1	2
−1	−1	1	−1	−1
−1	1	−1	−1	1
1	−1	−1	1	−1
1	1	1	1	1

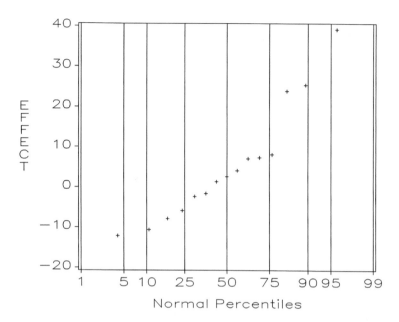

Fig. 5.28 This probability plot of effects estimates is used to visually check for effect significance. Effects that vary from the line determined by the residuals nearer the center of the plot are probably significant (4 = EENDP, 5 = POSTME, and the 34 = ETEMP*EENDP interaction). The PRESOL*ETEMP (13) interaction may also be significant because it falls below the fifth percentile of the distribution.

In fact, a 2^{3-1} experiment can collapse to one of three 2^2 experiments as shown in Figure 5.29.

In general, a 2^{k-p} fractional factorial collapses to a 2^q factorial, where q = k − p, if p factors are ignored. Thus a 2^{7-4} experiment collapses to a 2^3 factorial by ignoring four factors. It does not matter which four factors are ignored—any such arrangement will produce a 2^3 factorial in the remaining factors.

■ *Example 5.7: Collapsing a 2^{5-1} Experiment to a 2^4 Experiment*

In the case of the 2^{5-1} fractional factorial in Example 5.4, only one factor need be ignored to obtain a 2^4 factorial in the remaining factors. That factor was chosen by examining the ANOVA table and selecting the factor which seemed to have the least effect on the response on the basis of its sum of squares.

EGAS was selected. The analysis was redone with EGAS omitted:

Dependent Variable: CAP			
Source	DF	Sum of Squares	Pr > F
Model	10	12894.1250000	0.0022
Error	5	331.3125000	
Corrected total	15	13225.4375000	
Source	DF	Type III SS	Pr > F
PRESOL	1	248.06250000	0.1108
ETEMP	1	264.06250000	0.1024
PRESOL*ETEMP	1	588.06250000	0.0308
EENDP	1	6123.06250000	0.0002
PRESOL*EENDP	1	451.56250000	0.0476
ETEMP*EENDP	1	2575.56250000	0.0016
POSTME	1	2280.06250000	0.0020
PRESOL*POSTME	1	138.06250000	0.2085
ETEMP*POSTME	1	203.06250000	0.1404
EENDP*POSTME	1	22.56250000	0.5849

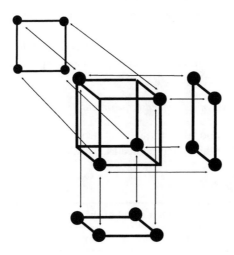

Fig. 5.29 The settings for a 2^{3-1} fractional factorial design are shown as points on a cube; any projection of the cube onto one of its defining planes produces a 2^2 factorial experiment.

Now the PRESOL*EENDP interaction appears significant; it did not seem so before the collapse. If EGAS has been wrongfully omitted, then the decision that PRESOL*EENDP is significant could be wrong.

While it might seem that collapsing could be employed to produce higher-resolution fractional factorials from low-resolution experiments, this does not work in many cases. A 2^{9-5} can sometimes be reduced to a 2^{7-3} by ignoring two factors, but a 2^{7-3} cannot be collapsed to a 2^{5-1}. This type of collapsing requires thoughtful planning, and previous knowledge of which factors will be dropped during analysis.

5.6 BLOCKING

Blocking is just as potent a design tool in fractional factorial experiments as in factorial experiments. As with factorials, fractional factorials can be run inside blocks, or split across blocks. By treating time as a blocking factor, fractional factorials can be also used sequentially to build knowledge while conserving experimental material.

5.6.1 Fractional Factorials in Blocks

Fractional factorials can be run within blocks if the number of runs required for a single replicate of the experiment is not larger than the block size. Any leftover runs in the block can be centerpoints, or they could be redundant runs for high-risk parts of the experiment.

Analysis of blocked fractional factorials proceeds as it did for factorial experiments. There is a slight loss of precision in blocked experiments when compared to experiments with true replication.

■ *Example 5.8: Analysis of a Blocked Experiment*

Suppose the 2^{3-1} fractional factorial experiment mentioned earlier had been run as three blocks rather then three replicates. The ANOVA would now have to account for the blocks:

	Dependent Variable: CAP			
Source	DF	Sum of Squares	F Value	Pr > F
Model	5	3092.16666667	7.64	0.0140
Error	6	485.83333333		
Corrected total	11	3578.00000000		
Source	DF	Type III SS	Pr > F	
PRESOL	1	560.33333333	0.0390	
PREDT	1	8.33333333	0.7592	
PRECHO	1	1452.00000000	0.0055	
BLOCK	2	1071.50000000	0.0304	

5.6.2 Fractional Factorials Split Across Blocks

Fractional factorials that are too large to fit into a single block can often be split across several blocks. In designing this type of experiment, the same principle applies to fractional factorials as was applied to factorials: *confound the highest-order interaction with the blocks.* As long as the number of blocks is a power of two, the assignment can be made by pretending that the blocks are extra factors, and constructing fractional factorials with these factors included in the experiment.

In simple cases, assignments of runs to blocks can be made in a manner which preserves the resolution of the original fractional factorial experiment, but there is no guarantee that this is always possible. Suppose that a 2^{5-2} fractional factor had to be split into two blocks. Call the blocks a sixth factor, and use the 2^{6-3} design. Both the 2^{5-2} and the 2^{6-3} are resolution III designs, so no resolution has been sacrificed by blocking.

■ *Example 5.9: Blocking and Resolution*

The layout for a 2^{5-2} (resolution III) design is shown below.

1	2	3	4	5
−1	−1	−1	1	1
−1	−1	1	1	−1
−1	1	−1	−1	1
−1	1	1	−1	−1
1	−1	−1	−1	−1
1	−1	1	−1	1
1	1	−1	1	−1
1	1	1	1	1

Suppose that this experiment had to be run in two blocks. Assign the blocks to factor 6, and use a 2^{6-3} design, which is still of resolution III:

1	2	3	4	5	Block
−1	−1	−1	1	1	1
−1	−1	1	1	−1	−1
−1	1	−1	−1	1	−1
−1	1	1	−1	−1	1
1	−1	−1	−1	−1	1
1	−1	1	−1	1	−1
1	1	−1	1	−1	−1
1	1	1	1	1	1

If the blocks could only accommodate two runs each, the experiment would have to be performed in four blocks. Use a 2^{7-4} (still resolution III) design by adding *two* additional factors, which decode into blocks as follows:

6	7	Block
−1	−1	1
−1	1	2
1	−1	3
1	1	4

The resulting design is shown here:

1	2	3	4	5	6	7	Block
−1	−1	−1	1	1	1	−1	3
−1	−1	1	1	−1	−1	1	2
−1	1	1	−1	1	−1	1	2
−1	1	1	−1	−1	1	−1	3
1	−1	−1	−1	−1	1	1	4
1	−1	1	−1	1	−1	−1	1
1	1	−1	1	−1	−1	−1	1
1	1	1	1	1	1	1	4

5.6.3   Augmented Designs

Augmentation is a sequential technique which enables the experimenter to increase the resolution of a fractional factorial by running another block of a *different but similar* fractional factorial. Augmentation is used when an initial experiment reveals that some factors have a significant effect on the response, but more information about factor effects must be obtained before process decisions can be made.

The main advantage of augmentation is that only a comparatively small amount of experimental material is committed to the first experiment—if factors are determined to be insignificant, then no further experiments with those factors need be run. The main disadvantage of augmentation is that the results of the first experiment must be obtained before starting the next experiment.

There are two common types of augmentation experiments. In the *foldover design,* a resolution IV design is produced from a resolution III design by changing the signs of all the experimental factors in the second experiment.

The second type of augmentation is used to resolve all (important) aliases for one particular factor of interest. This is done by changing signs *for that factor only* and leaving all other factor settings unchanged.

It is very likely that the artificially induced blocking factor in foldover designs (time) will be of practical importance, so it must be accounted for in any statistical analysis.

The procedure will be illustrated with a 2^{7-4} experiment folded over in each of the two ways mentioned above.

■ *Example 5.10: Augmenting a 2^{7-4} Experiment*

A 2^{7-4} fractional factorial experiment was conducted on the metal etch process with the following seven factors:
The experiment involved the seven factors:

- Preclean solution (PRESOL): water (–) or methanol (+)
- Preclean dip time (PREDT): 90 and 120 seconds
- Preclean contains choline (+) or not (–), (PRECHO)
- Etch gas 2 (EGAS): CCl_4 (–) or $CHCl_3$ (+)
- Etch temperature (ETEMP): 225 or 225°
- Etch endpoint method (EENDP): Al ion (–) or Zr ion (+)
- Postclean method (POSTME): multiple quench-rinse (–) or washer-drier (+)

The initial eight runs from a single replicate of this design are shown below:

PRESOL	PREDT	PRECHO	EGAS	ETEMP	EENDP	POSTME	CAP
−1	−1	−1	1	1	1	−1	210
−1	−1	1	1	−1	−1	1	252
−1	1	−1	−1	1	−1	1	209
−1	1	1	1	−1	1	−1	155
1	−1	−1	−1	−1	1	1	189
1	−1	1	−1	1	−1	−1	136
1	1	−1	1	−1	−1	−1	217
1	1	1	1	1	1	1	210

This design is of resolution III, so main effects are aliased with simple interactions. A resolution IV foldover design is obtained by changing the sign for every factor in the experiment, and repeating the experiment in another block (time). This was done with the results shown below:

PRESOL	PREDT	PRECHO	EGAS	ETEMP	EENDP	POSTME	CAP
1	1	1	−1	−1	−1	1	221
1	1	−1	−1	1	1	−1	179
1	−1	1	1	−1	1	−1	150
1	−1	−1	1	1	−1	1	167
−1	1	1	1	1	−1	−1	132
−1	1	−1	1	−1	1	1	218
−1	−1	1	−1	1	1	1	237
−1	−1	−1	−1	−1	−1	−1	234

Now, main effects are free of aliasing by interactions. Note in the analysis below that the BLOCK effect is included along with seven interactions of interest. Each of these interactions is aliased with two others, so care should be taken in interpreting the results.

Dependent Variable: CAP				
Source	DF	Sum of Squares	Pr > F	
Model	14	10818.0000000	0.08	0.9973
Error	1	10201.0000000		
Corrected total	15	21019.0000000		

Source	DF	Type III SS	Pr > F
BLOCK	1	100.00000000	0.9372
EGAS	1	1.00000000	0.9937
PRESOL	1	1980.25000000	0.7358
PREDT	1	72.25000000	0.9465
PRECHO	1	1056.25000000	0.8018
ETEMP	1	1521.00000000	0.7654
EENDP	1	25.00000000	0.9685
POSTME	1	5256.25000000	0.6036
PRESOL*EGAS	1	110.25000000	0.9341
PREDT*EGAS	1	56.25000000	0.9528
PRECHO*EGAS	1	2.25000000	0.9905
EGAS*ETEMP	1	400.00000000	0.8755
EGAS*EENDP	1	225.00000000	0.9061
EGAS*POSTME	1	12.25000000	0.9779

Assuming that the original eight runs had been obtained as before, an alternative experiment could have been conducted to break aliases with one factor of special interest. Suppose that POSTME was selected for further study; a second experiment would change signs for POSTME only, leaving all other factor levels unchanged:

PRESOL	PREDT	PRECHO	EGAS	ETEMP	EENDP	POSTME	CAP	BLOCK
−1	−1	−1	1	1	1	−1	210	1
−1	−1	1	1	−1	−1	1	252	1
−1	1	−1	−1	1	−1	1	209	1
−1	1	1	−1	−1	1	−1	155	1
1	−1	−1	−1	−1	1	1	189	1
1	−1	1	−1	1	−1	−1	136	1
1	1	−1	1	−1	−1	−1	217	1
1	1	1	1	1	1	1	210	1
−1	−1	−1	−1	1	1	−1	197	2
−1	−1	1	−1	−1	−1	1	207	2
−1	1	−1	1	1	−1	1	169	2
−1	1	1	1	−1	1	−1	123	2
1	−1	−1	1	−1	1	1	202	2
1	−1	1	1	1	−1	−1	149	2
1	1	−1	−1	−1	−1	−1	226	2
1	1	1	−1	1	1	1	233	2

Now both POSTME and all of its simple interactions can be tested:

Dependent Variable: CAP			
Source	DF	Sum of Squares	Pr > F
Model	14	8968.00000000	0.9990
Error	1	11025.00000000	
Corrected total	15	19993.00000000	

Source	DF	Type III SS	Pr > F
BLOCK	1	324.00000000	0.8919
PRESOL	1	100.00000000	0.9396
PREDT	1	0.00000000	1.0000
PRECHO	1	1482.25000000	0.7763
EGAS	1	25.00000000	0.9697
ETEMP	1	210.25000000	0.9126
EENDP	1	132.25000000	0.9306
POSTME	1	4160.25000000	0.6493
PRESOL*EGAS	1	4.00000000	0.9879
PREDT*EGAS	1	2209.00000000	0.7321
PRECHO*EGAS	1	42.25000000	0.9606
EGAS*ETEMP	1	182.25000000	0.9186
EGAS*EENDP	1	90.25000000	0.9426
EGAS*POSTME	1	6.25000000	0.9848

5.7   PLACKETT-BURMAN DESIGNS

The purpose of an initial screening design is to quickly determine which of many possible factors have a significant effect on a response, so designs of resolution III are usually acceptable choices. Plackett and Burman invented a type of resolution III experiment that requires only $4n$ runs for $4n - 1$ factors, where n is any positive integer. There is a Plackett-Burman design for 3 factors requiring 4 runs, a design for 7 factors requiring 8 runs, and a design for 11 factors in 12 runs.

These designs are very economical when compared with fractional factorials of the same resolution: the 2^{11-7} design is resolution III, but requires 16 runs; the corresponding Plackett-Burman design can be executed in 12 runs.

These very efficient designs do have some drawbacks:

- The effects of interactions cannot be estimated.
- Aliasing in Plackett-Burman designs is very complex compared to that in fractional factorials. In a Plackett-Burman design, every interaction is aliased to some extent with every main effect. If interactions are very large this can in rare cases cause some insignificant main effects to appear significant.

Plackett-Burman designs are constructed from tables, or with the aid of software. Several useful designs are shown in Table 5.2. Note that, with the exception of the last row, each column is simply a cyclic repeat of the previous column.

TABLE 5.2.   **Seven-Factor Plackett-Burman Design in Eight Runs**

1	2	3	4	5	6	7
1	-1	-1	-1	-1	-1	1
1	1	-1	-1	1	-1	1
1	1	1	-1	-1	1	-1
-1	1	1	1	-1	-1	1
1	-1	1	1	1	-1	-1
-1	1	-1	1	1	1	-1
-1	-1	1	-1	1	1	1
-1	-1	-1	-1	-1	-1	-1

1	2	3	4	5	6	7	8	9	A	B
1	-1	1	-1	-1	-1	1	1	1	-1	1
1	1	-1	1	-1	-1	-1	1	1	1	-1
-1	1	1	-1	1	-1	-1	-1	1	1	1
1	-1	1	1	-1	1	-1	-1	-1	1	1
1	1	-1	1	1	-1	1	-1	-1	-1	1
1	1	1	-1	1	1	-1	1	-1	-1	-1
-1	1	1	1	-1	1	1	-1	1	-1	-1
-1	-1	1	1	1	-1	1	1	-1	1	-1
-1	-1	-1	1	1	1	-1	1	1	-1	1
1	-1	-1	-1	1	1	1	-1	1	1	-1
-1	1	-1	-1	-1	1	1	1	-1	1	1
-1	-1	-1	-1	-1	-1	-1	-1	-1	-1	-1

TABLE 5.2 (continued).   15-Factor Plackett-Burman Design in 16 Runs

1	2	3	4	5	6	7	8	9	A	B	C	D	E	F
1	-1	-1	-1	1	-1	-1	1	1	-1	1	-1	1	1	1
1	1	-1	-1	-1	1	-1	-1	1	1	-1	1	-1	1	1
1	1	1	-1	-1	-1	1	-1	-1	1	1	-1	1	-1	1
1	1	1	1	-1	-1	-1	1	-1	-1	1	1	-1	1	-1
-1	1	1	1	1	-1	-1	-1	1	-1	-1	1	1	-1	1
1	-1	1	1	1	1	-1	-1	-1	1	-1	-1	1	1	-1
-1	1	-1	1	1	1	1	-1	-1	-1	1	-1	-1	1	1
1	-1	1	-1	1	1	1	1	-1	-1	-1	1	-1	-1	1
1	1	-1	1	-1	1	1	1	1	-1	-1	-1	1	-1	-1
-1	1	1	-1	1	-1	1	1	1	1	-1	-1	-1	1	-1
-1	-1	1	1	-1	1	-1	1	1	1	1	-1	-1	-1	1
1	-1	-1	1	1	-1	1	-1	1	1	1	1	-1	-1	-1
-1	1	-1	-1	1	1	-1	1	-1	1	1	1	1	-1	-1
-1	-1	1	-1	-1	1	1	-1	1	-1	1	1	1	1	-1
-1	-1	-1	1	-1	-1	1	1	-1	1	-1	1	1	1	1
-1	-1	-1	-1	-1	-1	-1	-1	-1	-1	-1	-1	-1	-1	-1

1	2	3	4	5	6	7	8	9	A	B	C	D	E	F	G	H	I	J
1	-1	1	1	-1	-1	-1	-1	1	-1	1	-1	1	1	1	1	-1	-1	1
1	1	-1	1	1	-1	-1	-1	-1	1	-1	1	-1	1	1	1	1	-1	-1
-1	1	1	-1	1	1	-1	-1	-1	-1	1	-1	1	-1	1	1	1	1	-1
-1	-1	1	1	-1	1	1	-1	-1	-1	-1	1	-1	1	-1	1	1	1	1
1	-1	-1	1	1	-1	1	1	-1	-1	-1	-1	1	-1	1	-1	1	1	1
1	1	-1	-1	1	1	-1	1	1	-1	-1	-1	-1	1	-1	1	-1	1	1
1	1	1	-1	-1	1	1	-1	1	1	-1	-1	-1	-1	1	-1	1	-1	1
1	1	1	1	-1	-1	1	1	-1	1	1	-1	-1	-1	-1	1	-1	1	-1
-1	1	1	1	1	-1	-1	1	1	-1	1	1	-1	-1	-1	-1	1	-1	1
1	-1	1	1	1	1	-1	-1	1	1	-1	1	1	-1	-1	-1	-1	1	-1
-1	1	-1	1	1	1	1	-1	-1	1	1	-1	1	1	-1	-1	-1	-1	1
1	-1	1	-1	1	1	1	1	-1	-1	1	1	-1	1	1	-1	-1	-1	-1
-1	1	-1	1	-1	1	1	1	1	-1	-1	1	1	-1	1	1	-1	-1	-1
-1	-1	1	-1	1	-1	1	1	1	1	-1	-1	1	1	-1	1	1	-1	-1
-1	-1	-1	1	-1	1	-1	1	1	1	1	-1	-1	1	1	-1	1	1	-1
-1	-1	-1	-1	1	-1	1	-1	1	1	1	1	-1	-1	1	1	-1	1	1
1	-1	-1	-1	-1	1	-1	1	-1	1	1	1	1	-1	-1	1	1	-1	1
1	1	-1	-1	-1	-1	1	-1	1	-1	1	1	1	1	-1	-1	1	1	-1
-1	1	1	-1	-1	-1	-1	1	-1	1	-1	1	1	1	1	-1	-1	1	1
-1	-1	-1	-1	-1	-1	-1	-1	-1	-1	-1	-1	-1	-1	-1	-1	-1	-1	-1

TABLE 5.2 (continued).   23-Factor Plackett-Burman Design in 24 Runs

1	*2*	*3*	*4*	*5*	*6*	*7*	*8*	*9*	*A*	*B*	*C*	*D*	*E*	*F*	*G*	*H*	*I*	*J*	*K*	*L*	*M*	*N*
1	-1	-1	-1	-1	1	-1	1	-1	-1	1	1	-1	-1	1	1	-1	1	-1	1	1	1	1
1	1	-1	-1	-1	-1	1	-1	1	-1	-1	1	1	-1	-1	1	1	-1	1	-1	1	1	1
1	1	1	-1	-1	-1	-1	1	-1	1	-1	-1	1	1	-1	-1	1	1	-1	1	-1	1	1
1	1	1	1	-1	-1	-1	-1	1	-1	1	-1	-1	1	1	-1	-1	1	1	-1	1	-1	1
1	1	1	1	1	-1	-1	-1	-1	1	-1	1	-1	-1	1	1	-1	-1	1	1	-1	1	-1
-1	1	1	1	1	1	-1	-1	-1	-1	1	-1	1	-1	-1	1	1	-1	-1	1	1	-1	1
1	-1	1	1	1	1	1	-1	-1	-1	-1	1	-1	1	-1	-1	1	1	-1	-1	1	1	-1
-1	1	-1	1	1	1	1	1	-1	-1	-1	-1	1	-1	1	-1	-1	1	1	-1	-1	1	1
1	-1	1	-1	1	1	1	1	1	-1	-1	-1	-1	1	-1	1	-1	-1	1	1	-1	-1	1
1	1	-1	1	-1	1	1	1	1	1	-1	-1	-1	-1	1	-1	1	-1	-1	1	1	-1	-1
-1	1	1	-1	1	-1	1	1	1	1	1	-1	-1	-1	-1	1	-1	1	-1	-1	1	1	-1
-1	-1	1	1	-1	1	-1	1	1	1	1	1	-1	-1	-1	-1	1	-1	1	-1	-1	1	1
1	-1	-1	1	1	-1	1	-1	1	1	1	1	1	-1	-1	-1	-1	1	-1	1	-1	-1	1
1	1	-1	-1	1	1	-1	1	-1	1	1	1	1	1	-1	-1	-1	-1	1	-1	1	-1	-1
-1	1	1	-1	-1	1	1	-1	1	-1	1	1	1	1	1	-1	-1	-1	-1	1	-1	1	-1
-1	-1	1	1	-1	-1	1	1	-1	1	-1	1	1	1	1	1	-1	-1	-1	-1	1	-1	1
1	-1	-1	1	1	-1	-1	1	1	-1	1	-1	1	1	1	1	1	-1	-1	-1	-1	1	-1
-1	1	-1	-1	1	1	-1	-1	1	1	-1	1	-1	1	1	1	1	1	-1	-1	-1	-1	1
1	-1	1	-1	-1	1	1	-1	-1	1	1	-1	1	-1	1	1	1	1	1	-1	-1	-1	-1
-1	1	-1	1	-1	-1	1	1	-1	-1	1	1	-1	1	-1	1	1	1	1	1	-1	-1	-1
-1	-1	1	-1	1	-1	-1	1	1	-1	-1	1	1	-1	1	-1	1	1	1	1	1	-1	-1
-1	-1	-1	1	-1	1	-1	-1	1	1	-1	-1	1	1	-1	1	-1	1	1	1	1	1	-1
-1	-1	-1	-1	1	-1	1	-1	-1	1	1	-1	-1	1	1	-1	1	-1	1	1	1	1	1
-1	-1	-1	-1	-1	-1	-1	-1	-1	-1	-1	-1	-1	-1	-1	-1	-1	-1	-1	-1	-1	-1	-1

*Layouts are given for Plackett-Burman designs with 7, 11, 15, 19, and 23 factors. Each experiment takes one more run than the number of factors, and aliasing is complex.

■ *Example 5.11: An 11-Factor Plackett-Burman Experiment*

A 12-run Plackett-Burman design was used to screen 11 metal etch process factors. Factors and their levels are:

- Preclean solution (PRESOL): water (–) or methanol (+)
- Preclean dip time (PREDT): 90 and 120 seconds
- Preclean contains choline (+) or not (–), (PRECHO)
- Etch oxide preetch time (EOXPE): 15 and 45 seconds (process target is 30 seconds)
- Etch gas 2 (EGAS): CCl_4 (–) or $CHCl_3$ (+)
- Etch temperature (ETEMP): 225 or 255°C
- Etch frequency (ECF): 11 MHz (–) or 11.5 MHz (+)
- Etch endpoint method (EENDP): Al ion (–) or Zr ion (+)
- Etch overetch time: 30 seconds (process target), and 40 seconds
- Postclean method (POSTME): multiple quench-rinse (–) or washer-drier (+)

- Postclean queue time (time from etch to postclean, POSTQT): 30 minutes (–) to 60 minutes (+)

The experiment was run in two identical blocks, thus providing an estimate for within-group variance needed for statistical tests. In this very complex experiment, replication also helps ensure that all runs are actually realized at least once. Run order within each block was randomized.

Raw data from the experiment is shown below:

B L O C K	P R E S O L	P R E D T	P R E C H O	E O X P E	E G A S	E T E M P	E C F	E N D P R	E O V E R	P O S T M E	P O S T Q T	C A P
1	1	-1	1	-1	-1	-1	1	1	1	-1	1	115
1	1	1	-1	1	-1	-1	-1	1	1	1	-1	160
1	-1	1	1	-1	1	-1	-1	-1	1	1	1	253
1	1	-1	1	1	-1	1	-1	-1	-1	1	1	121
1	1	1	-1	1	1	-1	1	-1	-1	-1	1	201
1	1	1	1	-1	1	1	-1	1	-1	-1	-1	151
1	-1	1	1	1	-1	1	1	-1	1	-1	-1	129
1	-1	-1	1	1	1	-1	1	1	-1	1	-1	173
1	-1	-1	-1	1	1	1	-1	1	1	-1	1	206
1	1	-1	-1	-1	1	1	1	-1	1	1	-1	182
1	-1	1	-1	-1	-1	1	1	1	-1	1	1	218
1	-1	-1	-1	-1	-1	-1	-1	-1	-1	-1	-1	229
2	1	-1	1	-1	-1	-1	1	1	1	-1	1	114
2	1	1	-1	1	-1	-1	-1	1	1	1	-1	214
2	-1	1	1	-1	1	-1	-1	-1	1	1	1	199
2	1	-1	1	1	-1	1	-1	-1	-1	1	1	175
2	1	1	-1	1	1	-1	1	-1	-1	-1	1	203
2	1	1	1	-1	1	1	-1	1	-1	-1	-1	154
2	-1	1	1	1	-1	1	1	-1	1	-1	-1	118
2	-1	-1	1	1	1	-1	1	1	-1	1	-1	172
2	-1	-1	-1	1	1	1	-1	1	1	-1	1	201
2	1	-1	-1	-1	1	1	1	-1	1	1	-1	161
2	-1	1	-1	-1	-1	1	1	1	-1	1	1	264
2	-1	-1	-1	-1	-1	-1	-1	-1	-1	-1	-1	201

An ANOVA is presented below. Because this is a resolution III design, no significance tests for interactions were requested.

Dependent Variable: CAP			
Source	DF	Sum of Squares	Pr > F
Model	12	35,135.6666667	0.0047
Error	11	6,064.8333333	
Corrected total	23	41,200.5000000	
Source	DF	Type III SS	Pr > F
BLOCK	1	60.1666667	0.7473
PRESOL	1	7,072.6666667	0.0043
PREDT	1	1,908.1666667	0.0898
PRECHO	1	13,348.1666667	0.0005
EOXPE	1	1,176.0000000	0.1721
EGAS	1	1,633.5000000	0.1132
ETEMP	1	988.1666667	0.2077
ECF	1	1,908.1666667	0.0898
EENDP	1	37.5000000	0.7991
EOVER	1	1,837.5000000	0.0952
POSTME	1	3,037.5000000	0.0387
POSTQT	1	2,128.1666667	0.0752

Thus, PRESOL, PRECHO and POSTME are all significant factors.

Even though Plackett-Burman designs are not fractional factorials, they can be folded over by switching all signs in the original design. This produces a resolution IV design to obtain estimates of the main factor effects which are not aliased with simple interactions.

5.8 SUMMARY

This chapter introduced two highly efficient types of screening design: 2^{k-p} fractional factorials and Plackett-Burman designs. Screening designs are usually the first to be executed when a new process is characterized, or when searching for the root cause of a problem in an established process. These designs can screen many factors quickly, but they do so at the expense of resolution. Hence, factorial experiments or resolution V fractional factorials usually follow.

Even within the screening segment of the experimentation sequence, it is prudent to run several experiments serially, rather than expend all available resources in one large experiment. Start with low-resolution experiments investigating many factors, and proceed to higher-resolution experiments on the few important factors.

Topics of importance not covered in this chapter include:

- 3^{k-p} fractional factorial experiments. These are very expensive experiments compared to the corresponding 2^{k-p} designs, but they do have the ability to estimate quadratic effects for quantitative factors. Response surface experiments (Chapter 6) are a much more cost-effective choice when estimating quadratic effects.

- D-optimal and other x-optimal designs. These designs exploit matrix algebra to maximize the sample F-value for particular types of alternatives to the null hypothesis. For example, D-optimal designs maximize a matrix determinant (hence the "D"). In some cases where experimental material is very expensive, and the probability of correctly completing every run of the experiment is high, these may be a reasonable choice. D-optimal designs can be produced only with the aid of suitable software.

CHAPTER 5 PROBLEMS

1. In which of the following situations would you use a 2^{11-7} fractional factorial experiment?

   (a) A brewer is trying to find out which of 11 trace minerals enhance the head on his favorite ale. He has no prior knowledge of their effects.

   (b) A plasma etch rate is known to be profoundly influenced by three factors: A, B, and C. Several other factors (D through K) have been suggested as possible influences on the etch rate, but no prior knowledge or scientific basis exists to support this supposition.

   (c) A potato chip manufacturer desires to determine which of 11 ingredients contribute to an "ideal crunch" experience. An ideal crunch is not really measurable, but specially trained tasters can tell if this occurs or if it does not.

   (d) A lithography coat process is in trouble—sometimes it produces a resist thickness within uniformity requirements (plus or minus 4% over the entire wafer), and sometimes produces a very nonuniform coating. 22 suspected influences have been selected for further study, and the problem solving team has ranked the influences by the order of their expected influence on uniformity.

   (e) A materials scientist experimenting with a novel niobium film capacitor has chosen 11 possible influences on capacitance as factors worthy of investigation: capacitor width, capacitor length, insulation thickness, and eight other similar parameters.

2. Give the layout for one replicate of a 2^{5-2} fractional factorial experiment with factors A, B, C, D, and E.

   (a) Are any of the main effects free of aliasing with some simple interaction?

   (b) Which simple interactions are aliased with one another?

   (c) What advantage does this type of experiment have over a 2^{5-1} fractional factorial experiment?

   (d) What is the resolution of the 2^{5-2} design?

3. Give the layout for one replicate of a 2^{6-3} fractional factorial design.

   (a) Are any of the main effects free of aliasing with some simple interaction?

   (b) Are any of the main effects aliased with one another?

   (c) Which simple interactions are aliased with one another?

   (d) Is any factor or main effect entirely free of important aliasing?

(*e*) What advantage does this type of experiment have over a 2^{5-2} fractional factorial experiment?

(*f*) What is the resolution of the 2^{6-3} design?

4. Give the layout for one replicate of a 2^{7-2} fractional factorial design with factors A, B, C, D, E, F, and G. Assign the factors in such a way that the EF interaction is free of important aliasing.

(*a*) Which main effects are free of aliasing with some simple interaction?

(*b*) Are any of the simple interactions aliased with one another?

(*c*) Which simple interactions are aliased with one another?

(*d*) What advantage does this type of experiment have over a 2^{7-3} fractional factorial experiment?

(*f*) When might a 2^{7-4} design be a better choice than this 2^{7-2}?

5. Give the layout for one replicate of a 2^{8-4} fractional factorial design with factors A, B, C, D, E, F, G, and H. Assign the factors in such a way that the EF interaction is aliased with the GH interaction.

(*a*) Are any main effects aliased with some simple interaction?

(*b*) Is any simple interaction free of important aliasing.

(*c*) What advantage does this type of experiment have over a 2^{8-3} fractional factorial experiment?

(*d*) When might a 2^{7-4} design be a better choice than this 2^{7-2} design?

6. Select a design to investigate the effect of seven factors on a response. This particular experiment should be able to estimate main effects free of important aliasing, but should be as economical as possible.

7. Select a design to investigate the effect of five factors on a response. This particular experiment should be able to estimate all simple interactions free of important aliasing.

8. Select a design to investigate the effect of 10 factors on a response. This particular experiment must allow estimation of the main effects free of important aliasing.

9. Select a design to investigate the effect of 11 factors on a response. This particular experiment must be run in less than 15 runs.

10. A microlithography engineer is investigating the effects of exposure energy, focus, postexposure bake (PEB) temperature, and PEB time on two responses of interest: critical dimension (CD) and the standard deviation of the CD. Exposure energy was varied about its target of 120 mJ in either direction; focus took the values 0 [microns (μm) out of the focus plane—this is perfect focus] and 5 microns (5 μm); PEB temperature was either 80 or 100°C; PEB time was 90 or 120 seconds. A 2^{4-1} fractional factorial experiment was performed in which each treatment combination was replicated three times. The results are shown below in standard order (even though the actual experiment was randomized).

EXP	FOC	Temp	Time		CD			STD CD	
110	0	80	90	3.44	3.46	3.49	0.020	0.018	0.018
110	0	100	120	3.40	3.46	3.46	0.014	0.020	0.01
110	5	80	120	3.73	4.03	3.71	0.050	0.060	0.05
110	5	100	90	3.82	3.76	3.75	0.064	0.068	0.056
130	0	80	120	3.18	3.19	3.27	0.014	0.010	0.01
130	0	100	90	3.19	3.22	3.18	0.012	0.030	0.022
130	5	80	90	3.31	3.42	3.40	0.050	0.052	0.054
130	5	100	120	3.74	3.52	3.67	0.54	0.052	0.052

(a) Do an ANOVA on the CD response. Are any of the factors significant? Is any simple interaction significant? What can you say about the significance of three-way interactions?

(b) Use graphical methods to check the validity of model assumptions. Are any assumptions violated?

(c) Compute the standard deviation for each treatment combination, and takes its square root to normalize it. Determine point estimates for the effects of the factors on standard deviation. Do any of the factors appear to be significant?

(d) Draw an interaction plot depicting the effects of focus and energy on the square root of the standard deviation.

11. Lactobacillus bacteria produce lactic acid when introduced into a suitable medium and allowed to grow. A chemical manufacturer ran an experiment to investigate the effect of nine different trace ions on lactic acid formation. The ions were assumed not to interact with one another, so a 2^{9-5} experiment was used. Because this type of reaction is prone to contamination by other bacteria, the entire experiment was replicated four times. Results are shown below.

Na	Ca	Mg	HSO_3	BO_4	PO_4	Fe	K	Mn		Lactic Acid (mL)		
−1	−1	−1	−1	−1	−1	−1	−1	1	17.7	20	18.4	19.8
−1	−1	−1	1	−1	1	1	1	−1	9.8	6.2	5.9	6.6
−1	−1	1	−1	1	1	1	−1	−1	24.5	22	24.2	24.8
−1	−1	1	1	1	−1	−1	1	1	0	0	0.3	0
−1	1	−1	−1	1	1	−1	1	−1	23.9	24.3	24.7	26.2
−1	1	−1	1	1	−1	1	−1	1	0	0.5	0	0
−1	1	1	−1	−1	−1	1	1	1	17	19.4	17.6	18.9
−1	1	1	1	−1	1	−1	−1	−1	11	7.1	7.7	7
1	−1	−1	−1	1	−1	1	1	−1	18.9	20.8	17.1	15.9
1	−1	−1	1	1	1	−1	−1	1	9.8	11.3	10.6	11.8
1	−1	1	−1	−1	1	−1	1	1	30.4	29.8	33	29.7
1	−1	1	1	−1	−1	1	−1	−1	2.3	2.9	3.5	2.8
1	1	−1	−1	−1	1	1	−1	1	32.3	31.1	31.9	30
1	1	−1	1	−1	−1	−1	1	−1	4.7	0.4	0.7	0
1	1	1	−1	1	−1	−1	−1	−1	19.4	18.3	15.5	18.1
1	1	1	1	1	1	1	1	1	9.7	11.3	12.6	13.1

(*a*) Do an ANOVA on the lactic acid concentration.

(*b*) Would it have been better to do a twice-replicated 2^{9-4} experiment rather than four replicates of the 2^{9-5} experiment?

(*c*) Would it have been better to do a single 2^{9-3} experiment rather than the replicated 2^{9-5} experiment?

(*d*) What factors should influence your choice of experiment type in this context?

12. A metrology engineer is interested in determining how the seven settings on her new scanning electron microscope affect measurement speed, as measured by the time it takes to locate and measure the 15 standard patterns on a test wafer. She used an unreplicated 2^{7-3} fractional experiment, and added six centerpoints so statistical tests could be done. Results are shown below:

S1	S2	S3	S4	S5	S6	S7	Time	CD STD
−1	−1	−1	−1	−1	−1	−1	88	0.00394
−1	−1	−1	1	−1	1	1	79	0.0039
−1	−1	1	−1	1	1	1	89	0.0023
−1	−1	1	1	1	−1	−1	82	0.00247
−1	1	−1	−1	1	1	−1	90	0.00315
−1	1	−1	1	1	−1	1	101	0.0037
−1	1	1	−1	−1	−1	1	103	0.00264
−1	1	1	1	−1	1	−1	102	0.00237
1	−1	−1	−1	1	−1	1	79	0.00362
1	−1	−1	1	1	1	−1	71	0.00398
1	−1	1	−1	−1	1	−1	72	0.00314
1	−1	1	1	−1	−1	1	84	0.00249
1	1	−1	−1	−1	1	1	74	0.00384
1	1	−1	1	−1	−1	−1	89	0.00372
1	1	1	−1	1	−1	−1	108	0.00246
1	1	1	1	1	1	1	81	0.00255
0	0	0	0	0	0	0	85	0.0031
0	0	0	0	0	0	0	81	0.00267
0	0	0	0	0	0	0	75	0.00273
0	0	0	0	0	0	0	73	0.00289

(*a*) Do an ANOVA to determine which of the seven settings have a significant influence on measurement time.

(*b*) Describe which main effects and simple interactions are aliased with one another.

(*c*) Pick one main effect which appears to be significant, and explain the consequences of aliasing when interpreting this apparent significance.

13. The metrology engineer in the previous problem is also interested in the uniformity of the measurements obtained, so she collected data on the SD of the 15 test sites each time the test wafer was measured. This data was shown in the CD STD column of the table above.

(*a*) Do an analysis of variance with the square root of the standard deviation as a response.

(*b*) Test the ANOVA assumptions; is this a valid model?

(*c*) Is the speed of a measurement related in any way to the standard deviation? Quantify this relationship.

14. A brewer is experimenting with the recipe for Grizzly Beer Ale as explained in Papazian (1991). The standard recipe calls for (among other things), 1 lb of crystal malt, 2 tsp gypsum, 1.5 oz Fuggles hops (for boiling), and 0.5 oz Hallertauer hops (for finishing). In addition to varying these factors, the addition of sodium chloride to the wort was considered, as was the addition of 1/4 oz of Orion hops (for boiling). The factors in this 2^{6-3} fractional factorial experiment and their settings are shown below. Four centerpoints were also run; the "0" setting is the centerpoint.

	Settings		
Factor	*−1*	*0*	*+1*
Crystal malt	0.95 lb	1 lb	1.05 lb
Gypsum	1 tsp	2 tsp	3 tsp
uggles hops	1.0 oz	1.5 oz.	2.0 oz
Hallertauer	0.0 oz	0.5 oz.	1.0 oz
NaCl	0	0	3 tsp
Orion Hops	0.0 oz	0	0.5 oz

The beer was brewed, tasted, and measured on the standard 50-point beer tasting scale. Results are shown below:

Malt	*Gypsum*	*F.Hops*	*H.Hops*	*NaCl*	*O.Hops*	*Score*
−1	−1	−1	1	1	1	4
−1	−1	1	1	−1	−1	37
−1	1	−1	−1	1	−1	6
−1	1	1	−1	−1	1	28
1	−1	−1	−1	−1	1	42
1	−1	1	−1	1	−1	1
1	1	−1	1	−1	−1	41
1	1	1	1	1	1	2
0	0	0	0	0	0	36
0	0	0	0	0	0	30
0	0	0	0	0	0	36
0	0	0	0	0	0	37

(a) Do an analysis of variance to determine the effect of the factors on taste. Make interaction plots for any significant interactions.

(b) Do you think that the choice of factor settings was effective in this case?

(c) What recipe would you recommend on the basis of this data.

15.   A tennis player with a rather weak serve has decided to use statistics to improve his serve speed. He conducted an unreplicated 2^{5-1} fractional factorial experiment with the following factors and settings:

Factor	Settings	
	−1	+1
Head position during toss (Head)	Service line on opponent's court	In the air, at peak of balltoss
Back posture (Back)	Straight	Arched back
Right foot alignment (Foot)	Parallel to the baseline	60 degree angle from baseline
Hydration	No water before serving	Drink 50 ml of water before each serve
Breathing (Breath)	Inhale while striking	Exhale while striking

Results from the experiment are shown below:

Head	Back	Foot	Hydr.	Breath	Speed
−1	−1	−1	−1	1	32
−1	−1	−1	1	−1	37
−1	−1	1	−1	−1	31
−1	−1	1	1	1	40
−1	1	−1	−1	−1	42
−1	1	−1	1	1	32
−1	1	1	−1	1	31
−1	1	1	1	−1	29
1	−1	−1	−1	−1	47
1	−1	−1	1	1	41
1	−1	1	−1	1	50
1	−1	1	1	−1	48
1	1	−1	−1	1	61
1	1	−1	1	−1	57
1	1	1	−1	−1	66
1	1	1	1	1	53

(a) Estimate the effects and their simple interactions, and use a normal probability plot to make decisions regarding their significance.

(b) What recommendations would you make to this tennis player?

(*c*) Assume for a moment that the hydration factor was known to have an insignificant effect on the response, and use ANOVA to estimate the effects of the remaining factors and all of their interactions. Are your conclusions the same as those in (a)?

(*d*) How would you have designed this experiment if the tennis player could only serve eight serves at a time without fatigue?

16. Six factors were varied in a 2^{6-3} fractional factorial experiment as follows:

A	B	C	D	E	F
−1	−1	−1	1	1	1
−1	−1	1	1	−1	−1
−1	1	−1	−1	1	−1
−1	1	1	−1	−1	1
1	−1	−1	−1	−1	1
1	−1	1	−1	1	−1
1	1	−1	1	−1	−1
1	1	1	1	1	1

(*a*) List the eight additional treatment combinations of a foldover design which could transform this into a resolution IV experiment.

(*b*) List the treatment combinations which, if added to the original 2^{6-3} design, would enable you to estimate the effects of factor C and all of its simple interactions free of important aliasing.

17. Ten factors (1 through A) were varied in a 2^{10-6} fractional factorial experiment as shown in Table 5.1.

(*a*) List the eight additional treatment combinations of a foldover design that could transform this into a resolution IV experiment.

(*b*) List the treatment combinations which, if added to the original 2^{10-6} design, would enable you to estimate the effects of factor "5" and all of its simple interactions free of important aliasing.

18. Ten factors ("1" through "A") are to be involved in a 2^{10-5} fractional factorial experiment.

(*a*) List the treatment combinations that will be used in this experiment.

(*b*) Suppose the experiment must be split across two blocks of size 16. Assign treatment combinations to blocks in a way that still allows for the estimation of interesting effects.

(*c*) Suppose the experiment must be split across four blocks of size eight. Assign treatment combinations to blocks in a way that still allows for the estimation of interesting effects.

19. A bowler has decided to investigate the effects of seven factors on his bowling scores. The factors and their settings are shown below:

	Settings	
Factor	*−1*	*+1*
Shoes	Bowling shoes	Cowboy boots
Glove	No glove	Wear the glove
Sighting point	Sight down	Sight toward pins
Ball weight	14 lb	16 lb
Glasses	Bifocals	Sport glasses (long-range vision optimized)
Starting position	Third dot	Fourth dot
Starting foot	Left	Right

A 2^{7-2} fractional factorial experiment was conducted over the course of four evenings, with the results shown below:

Day	Shoe	Glv	Sgt	Ball	Gls	Pos	Foot	Score
4	−1	−1	−1	−1	−1	1	1	268
1	−1	−1	−1	−1	1	−1	−1	149
2	−1	−1	−1	1	−1	−1	−1	137
3	−1	−1	−1	1	1	1	1	256
3	−1	−1	1	−1	−1	−1	−1	153
2	−1	−1	1	−1	1	1	1	272
1	−1	−1	1	1	−1	1	1	260
4	−1	−1	1	1	1	−1	−1	141
1	−1	1	−1	−1	−1	−1	1	188
4	−1	1	−1	−1	1	1	−1	220
3	−1	1	−1	1	−1	1	−1	208
2	−1	1	−1	1	1	−1	1	176
2	−1	1	1	−1	−1	1	−1	224
3	−1	1	1	−1	1	−1	1	192
4	−1	1	1	1	−1	−1	1	180
1	−1	1	1	1	1	1	−1	212
1	1	−1	−1	−1	−1	1	−1	220
4	1	−1	−1	−1	1	−1	1	188
3	1	−1	−1	1	−1	−1	1	176
2	1	−1	−1	1	1	1	−1	208
2	1	−1	1	−1	−1	−1	1	192
3	1	−1	1	−1	1	1	−1	224
4	1	−1	1	1	−1	1	−1	212
1	1	−1	1	1	1	−1	1	180
4	1	1	−1	−1	−1	−1	−1	149
1	1	1	−1	−1	1	1	1	268
2	1	1	−1	1	−1	1	1	256
3	1	1	−1	1	1	−1	−1	137
3	1	1	1	−1	−1	1	1	272
2	1	1	1	−1	1	−1	−1	153
1	1	1	1	1	−1	−1	−1	141
4	1	1	1	1	1	1	1	260

(a) Treat the day as a blocking factor, and obtain point estimates of factor effects.

(b) Does the day (block) make a difference? What results would be obtained if the day were ignored?

(c) Use a normal probability plot to decide which main effects are significant.

(d) Suppose that one more night of bowling (eight games) could be arranged, and that the interaction of the Glove and Glasses factors were of interest. What treatment combinations should be tried on this last night?

(e) What advice would you give this bowler?

20. A metrology engineer is interested in determining how the seven settings on her new scanning electron microscope affect measurement speed, as measured by the time it takes to locate and measure the 15 standard patterns on a test wafer. She used an unreplicated Plackett-Burman experiment with eight runs, and added six centerpoints so statistical tests could be done. Results are shown below:

A	B	C	D	E	F	G	Time	CD Std
1	−1	−1	1	−1	1	1	65	0.00394
1	1	−1	−1	1	−1	1	91	0.0039
1	1	1	−1	−1	1	−1	91	0.0023
−1	1	1	1	−1	−1	1	91	0.00247
1	−1	1	1	1	−1	−1	85	0.00236
−1	1	−1	1	1	1	−1	89	0.0037
−1	−1	1	−1	1	1	1	78	0.00264
−1	−1	−1	−1	−1	−1	−1	100	0.00316
0	0	0	0	0	0	0	99	0.00323
0	0	0	0	0	0	0	102	0.00366
0	0	0	0	0	0	0	103	0.00415
0	0	0	0	0	0	0	104	0.00388
0	0	0	0	0	0	0	99	0.00417
0	0	0	0	0	0	0	97	0.00434

(a) Do an analysis of variance to determine which of the seven settings have a significant influence on measurement time.

(b) Describe which main effects and simple interactions are aliased with one another.

(c) Pick one main effect that appears to be significant, and explain the consequences of aliasing when interpreting this apparent significance.

(d) Give a set of eight more treatment combinations that would turn this resolution III design into a resolution IV design. (Ignore the centerpoints.)

21. A lithography coat process is in trouble; sometimes it produces a resist thickness within uniformity requirements (plus or minus 4% over the entire wafer), and sometimes produces a very nonuniform coating. Uniformity in this context is the resist thickness standard deviation divided by the wafer average, multiplied by 100:

$$\text{Unif} = 100 \ \frac{S_{\text{Thick}}}{\overline{X}}$$

Eleven suspected influences were selected for further study, and a 12-run Plackett-Burman design with four centerpoints was used. The results of the experiment are shown below:

A	B	C	D	E	F	G	H	J	K	L	Unif.
1	−1	1	−1	−1	−1	1	1	1	−1	1	3.19
1	1	−1	1	−1	−1	−1	1	1	1	−1	4.53
−1	1	1	−1	1	−1	−1	−1	1	1	1	5.40
1	−1	1	1	−1	1	−1	−1	−1	1	1	1.41
1	1	−1	1	1	−1	1	−1	−1	−1	1	5.24
1	1	1	−1	1	1	−1	1	−1	−1	−1	0.09
−1	1	1	1	−1	1	1	−1	1	−1	−1	0.41
−1	−1	1	1	1	−1	1	1	−1	1	−1	3.29
−1	−1	−1	1	1	1	−1	1	1	−1	1	0.43
1	−1	−1	−1	1	1	1	−1	1	1	−1	0.50
−1	1	−1	−1	−1	1	1	1	−1	1	1	0.49
−1	−1	−1	−1	−1	−1	−1	−1	−1	−1	−1	4.81
0	0	0	0	0	0	0	0	0	0	0	2.24
0	0	0	0	0	0	0	0	0	0	0	2.21
0	0	0	0	0	0	0	0	0	0	0	3.60
0	0	0	0	0	0	0	0	0	0	0	1.09

(a) Use analysis of variance to test the significance of the 11 factors.
(b) What conclusions can you make about the significance of simple interactions?
(c) Suppose that main effects estimates free of simple interaction aliasing were desired. What additional 12 runs could be used to provide these estimates?
(d) Suppose instead that three factors appeared to have a significant effect on the resist uniformity. What experiment would you suggest to further investigate these factors and their interactions?

CHAPTER 6

OPTIMIZATION METHODS

6.1 INTRODUCTION

Optimization methods are applied to discover ideal target combinations for quantitative factors. This should be done after screening experiments have sorted through the many possible factors which influence important responses and selected the critical few that must be carefully targeted and controlled to produce consistently optimum responses.

Optimization experiments are more complex than screening experiments; because optimization experiments must be able to make rather accurate estimates of the effects of factors and their interactions on responses, they usually require more than two levels per factor. Most optimization experiments in actual practice involve only two or three factors, so that this additional complexity can be accommodated within experiments of a reasonable size.

This chapter introduces two very different optimization techniques: the direction of steepest ascent (DSA) and the simplex method. Simplex designs (Section 6.4) require only one additional run per iteration (after an initial larger experiment), so they are very efficient when experimental runs take little time to perform and analyze. The DSA method (Section 6.3) is an application of linear regression which makes uses carefully supplemented factorials to point the experimenter in a direction that will discover desirable setting combinations with a few large experiments.

Section 6.2 prefaces the presentation of these designs and their applications with an introduction to response surfaces and some common methods of visualizing them.

6.2   RESPONSE SURFACES

There are many physical situations in which the value one variable essentially predicts the value of another. For example, the relationship between the distance a particular rock has fallen and the time since it was dropped is well known, and can be represented with an equation:

$$d = \frac{1}{2} at^2$$

where a is a constant. This relationship is shown visually in Figure 6.1.

The time since the rock was dropped is a *predictor,* and the distance fallen in that time is a *response.* The equation and the plot are descriptions of the *response surface* that results when time is used to predict distance fallen.

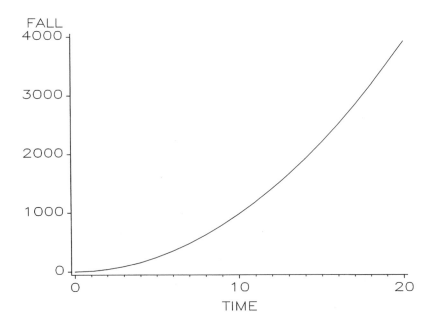

Fig. 6.1   The relationship between the distance a rock has fallen and the time since it was dropped represented here. The predictor is the time since the rock was dropped, and is shown on the x-axis. The response is the distance fallen in that time, and is shown on the y-axis.

In real life, the actual response surface relating predictors to responses is never known. Instead, experiments and scientific knowledge must be used to estimate the surface.

The concept of a response surface easily generalizes to physical situations with more than one predictor. A response surface taking into account air resistance of an object dropped as well as time since the drop is an actual surface in three-dimensional space as shown in Figure 6.2. The two predictors are the time since dropping and wind resistance; the response is the distance fallen.

A more convenient way of visualizing response surfaces is with contour plots. This is actually a plot of level curves, but the "contour plot" nomenclature is so universally misused that the error is deliberately perpetuated in this book. See p. 862 of Thomas and Finney (1988) for an explanation of the difference. A contour plot is a set of labeled curves representing a line of constant response, much like a map used to plan hikes (Figure 6.3). Contour plots are produced by most statistical software, and they have definite advantages over three-dimensional response surface plots:

- The are less prone to misinterpretation caused by differences in scaling or rotation.
- They facilitate the plotting of experimental settings—this is much more difficult to do under a three-dimensional surface.
- They provide more intuition regarding the choice of optimum settings or the direction in which to move for the next experiment, because response values can be unambiguously marked on the contour plot.

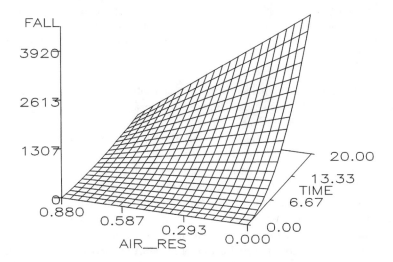

Figure 6.2 Two predictors and their response define a response surface. The time since dropping a rock and the air resistance are plotted on the *x*- and *y*-axes in the horizontal plane. The response— distance fallen—is plotted on the *z*-axis.

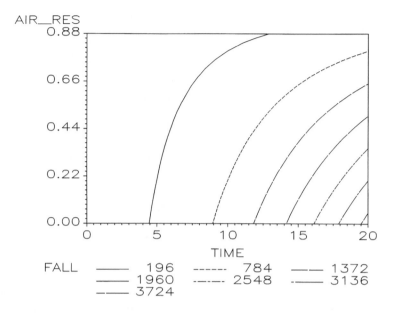

Fig. 6.3   This contour plot is a set of labeled curves representing a line of constant response—distance fallen in this case. The predictors are shown on the *x*- and *y*-axes of the plot. The legend below the plot is necessary to interpret the plot.

- By overlaying contour plots for two responses (Figure 6.4) the merits of proposed targets can be considered with respect to both responses simultaneously. This is an essential capability in cases where some tradeoff decisions regarding responses must be made.

There are some archetypal surfaces that occur often in practice: the stationary ridge (Figure 6.5), the rising ridge (Figure 6.6), the simple maximum (Figure 6.7), and the saddle point (Figure 6.8).

Real response surfaces can vary widely from these simple forms—they often have cliffs (Figure 6.9), or multiple optima (Figure 6.10).

Response surfaces also exist in situations with more than two predictors, but they are difficult to visualize. For this reason that direction of steepest ascent (Section 6.3) is first developed thoroughly for the two-predictor case. In this way the reader can thoroughly exercise intuition where it is most easily applied, much as a beginning pilot learns to fly in daylight and good weather.

Section 6.3.4 develops the application of the DSA method to response surfaces with more than two predictors. Visualization in this situation can be quite difficult so this case is much like flying an aircraft by instruments alone: ample faith in the method and skill in its application are required.

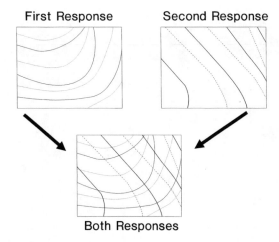

Fig. 6.4 Contour plots for two individual responses with the same predictors are shown in the top of the figure; these are superimposed in the bottom figure. If contour lines are well coded and labeled, the merits of proposed targets can be considered with respect to both responses simultaneously.

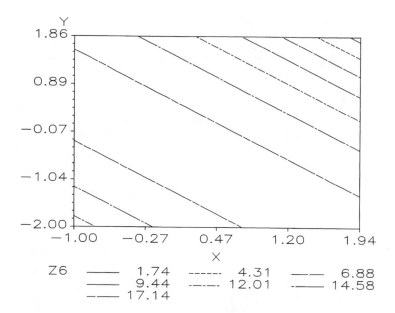

Fig. 6.5 The stationary ridge response surface is often seen in practice. In this case, the axis of the ridge is set at an angle to the x-axis, and changes in the response are most quickly obtained by changing the predictor on the y-axis.

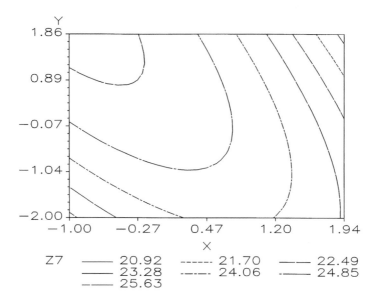

Fig. 6.6   This rising ridge has its maximum at the upper left of the contour plot, and a minimum at the upper right.

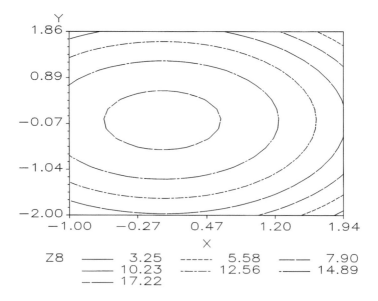

Fig. 6.7   The simple maximum response surface comes to a single largest response value within the range of the predictors.

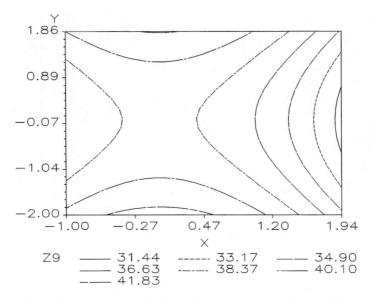

Fig. 6.8   This saddle point response surface has high values at the top and bottom of the graph, and lower values at the right and left of the graph.

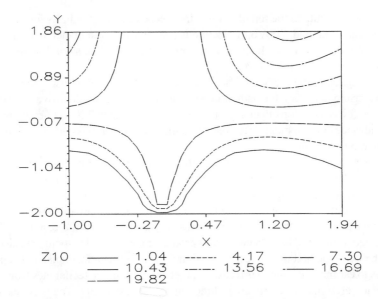

Fig. 6.9   This response surface has cliffs—sudden changes observed when a particular combination of predictors is encountered.

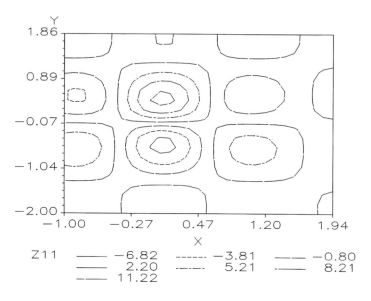

Fig. 6.10   This response surface has multiple optima—both maxima and minima.

6.3   DIRECTION OF STEEPEST ASCENT

The quickest route to the top of a hill is the steepest; this is the basis for the DSA optimization method. DSA optimization is a stepwise method wherein every step taken is in the direction which, based on present knowledge, should increase the response the most for that given step size.

The appeal of the method of steepest ascent is its efficiency: because it points the experimenter along the path of most rapid increase, it should arrive at an optimum in fewer iterations than any other method. This appeal is augmented by recent advances in statistical software that facilitate computations and the visual representation of results.

Linear regression is a tool to estimate the nature of the relationship between predictors and a response. The results of regression analysis are a decision regarding the statistical significance of each of the predictors, a measure of the amount of variance of the response accounted for by the prediction equation (R^2), and an equation describing the relationship between predictors and the response. That same equation could have been used to produce an estimated response surface.

The DSA method uses linear regression analysis of data from planned experiments to estimate response surfaces, and then applies simple techniques from calculus to determine the (estimated) direction of steepest ascent. The only real difference between regression analyses and those used for DSA is in the design of the experiment: factor settings are carefully chosen so that the DSA can be estimated with a relatively small number of experimental runs. The designs also take into account the supposed form of the functional relationship between predictors and response (linear, quadratic, etc.), and provide some means of checking the applicability of the model. The DSA experimenter can estimate the response surface while enjoying the efficiency and risk management of a planned experiment.

The language of optimization experiments has been strongly influenced by its regression heritage: experimental factors are called *predictors* or *independent variables,* and responses are called *dependent variables.*

The veracity of predictions based on an estimated response surface depend on the appropriateness (fit) of the chosen model, and on the sampling error to which every experiment is subject. The adverse effects of sampling error can be remedied with sufficient sample size and good experimental practices like randomization and blocking, Fitting the wrong model is a more serious problem which can result in completely wrong conclusions.

For example, suppose the curved surface in Figure 6.10 were wrongly fitted with a linear model without interaction:

$$z_i = \beta_0 + \beta_1 x_i + \beta_2 y_i + \varepsilon_i$$

Recall that:

- β_0 is the true intercept (any estimate of this population parameter will be called b_0),
- β_1 is the true coefficient of the first predictor variable, x. x_i is one observation of this variable.
- β_2 is the true coefficient of the second predictor variable, y. y_i is one observation of this variable.
- and ε_i is one sample from a population of independent, normally distributed residuals with constant variance and zero mean.

The surface predicted by such a model is shown in Figure 6.11. Any conclusions regarding the surface is suspect, and any extrapolation away from the original data is likely to be very misleading.

If a quadratic model with interaction were fit to this same surface:

$$z_i = \beta_0 + \beta_1 x_i + \beta_2 y_i + \beta_3 x_i y_i + \beta_4 x_i^2 + \beta_5 y_i^2 + \varepsilon_i$$

then the predicted response surface (Figure 6.12) would provide more accurate information about the actual surface, at least in the neighborhood in which measurements were made.

Sometimes the physical setting can give clues to the correct form of model, as in the falling rock example. The more common situation is that almost nothing is known about the shape of the surface, and some reasonable guesses must be made.

Since DSA experiments are designed to check model fit, there is little harm in choosing a very simple model initially. If this simple model is found to be inappropriate, it can be augmented with additional runs or a more fitting model can be used for the next experiment. Because simple surfaces are cheaper to estimate than

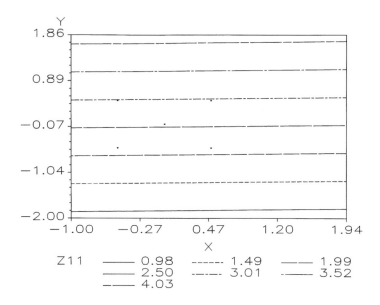

Fig. 6.11   A linear model without interaction was used to model the complex surface shown in Fig. 6.10. The points in a factorial experiment with centerpoints (marked on the plot) were used to estimate the parameters of the model. This is a poor fit, and would be entirely unsuitable for prediction on this surface.

complex surfaces (they require fewer runs per experiment), this strategy has the advantage that a minimum of experimental material is required overall.

One way to realize this cost-effective experimental strategy is to begin with factorial experiments with centerpoints, and then advance to the more elaborate central composite design when the simpler factorial models exhibit lack of fit.

6.3.1   Factorial Designs for DSA

Factorial designs for DSA optimization are very similar to factorials used for other purposes. Centerpoints are added to the optimization designs so that the fit of the model can be assessed, and to provide an estimate of within-group variance if the remainder of the factorial is unreplicated. For most experiments, the center-point of the factorial is chosen to be the best guess at the optimum setting.

If x and y are predictors, and z is a response, then factorial designs can estimate a model of the form:

$$z_i = \beta_0 + \beta_1 x_i + \beta_2 y_i + \beta_3 x_i y_i + \varepsilon_i$$

Here, $\beta_0, \beta_1, \beta_2,$ and β_3 are population parameters. They will be estimated with linear regression, and estimates will be named $b_0, b_1, b_2,$ and b_3. The interaction term in this context is actually a product term.

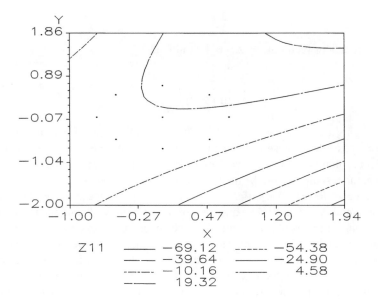

Fig. 6.12   A quadratic model with interaction was fit to the surface in Fig. 6.10; the points of the (central composite) experiment used to estimate the parameters of the model are plotted on the graph. This model will provide more accurate information about the actual surface, at least in the neighborhood in which measurements were made.

Once model coefficients are estimated, finding the direction of steepest ascent is simply a matter of arithmetic. The calculus immediately below is necessary to derive these arithmetic formulae, but it is not necessary to understand and use the designs.

Recall from multivariable calculus that a function in several variables will increase most rapidly if changed in the direction of the *gradient*. The gradient of a two-predictor function at any point is:

$$\nabla (x, y) = \begin{pmatrix} \dfrac{\partial z}{\partial x} \\[2mm] \dfrac{\partial z}{\partial y} \end{pmatrix}$$

In the case of the simple function above this equation simplifies to:

$$\nabla (x, y) = \begin{pmatrix} b_1 + b_3 \, y \\[1mm] b_2 + b_3 \, x \end{pmatrix}$$

So if z were thought to be predicted by x and y as follows:

$$z = 13 + 8x - 10y + 2xy$$

then the gradient vector at $x = 2$, $y = 4$ would be:

$$\begin{pmatrix} 8 + 2\,(4) \\ -10 + 2\,(2) \end{pmatrix} = \begin{pmatrix} 16 \\ -6 \end{pmatrix}$$

Changing x and y in proportion to the terms of this vector should result in the quickest possible increase in z. This is verified visually in Figure 6.13 where the gradient is plotted on the response surface for z. By proceeding in the opposite direction, the response would be made to decrease most rapidly.

As long as the response surface has the same simple shape predicted by this regression model, changes in the direction of the gradient will continue to increase the response. In most physical situations the shape of the response surface is too complex to predict with a simple factorial, so a sequence of experiments is used to periodically reassess the model and recompute the direction of steepest ascent.

The centerpoint for the next experiment in the sequence will be somewhere along the gradient vector from the last centerpoint, and at a reasonable distance from that centerpoint. If the new centerpoint is too far from the old, the surface may change so radically that the new experiment is of little value. If the new centerpoint to too close to the old, experimental material will be wasted because much of the new experiment is redundant with the previous experiment.

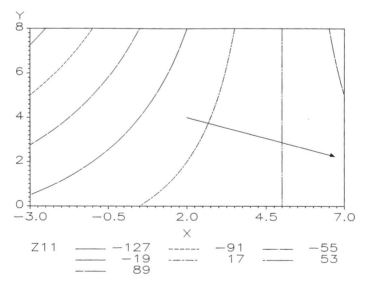

Fig. 6.13   The direction of steepest ascent from a particular point is in the direction of the gradient of the function describing the response surface.

The choice of this *step size* is largely dependent on the situation. In cases where the physics or chemistry of the process are well understood and the surface is known to be somewhat smooth, a larger step size may be an effective choice. In complex or poorly understood surfaces, smaller step sizes are better. In the absence of any previous knowledge about the surface, a rule of thumb is to make the step size about equal to the distance of the maximum predictor setting from the factorial centerpoint as in Figure 6.14. In this way, some overlap between the old and new experiments is always achieved.

Assuming that a gradient has been determined and a step size chosen, the centerpoint for the next experiment can be found by advancing the step size along the gradient vector from the centerpoint of the previous experiment.

The worksheet below outlines the computation. Denote the old centerpoint coordinates as x_{cp} and y_{cp}. Call the step size "Step," and write the estimated equation for the response surface as follows:

$$z = b_0 + b_1 x + b_2 y + b_3 xy$$

Follow the instructions in the "Action" column to arrive at the results in the "Result" column; the last two results give the x and y coordinates of the next centerpoint.

Step	Action	Result
1	$R1 = b_1 + b_3 y_{cp}$	R1
2	$R2 = b_2 + b_3 x_{cp}$	R2
3	$R3 = \sqrt{(R1)^2 + (R2)^2}$	R3
4	$R4 = \text{Step} \dfrac{R1}{R3}$	R4
5	$R5 = \text{Step} \dfrac{R2}{R3}$	R5
6	New $x_{cp} = x_{cp} + R4$	new x_{cp}
7	New $y_{cp} = y_{cp} + R5$	new y_{cp}

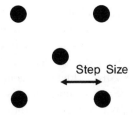

Step Size

Fig. 6.14   The design points of a 2^2 factorial experiment with centerpoints are shown. If the step size is chosen to be the difference from a centerpoint to the side of the experiment, sufficient overlap between experiments will occur to prevent surprises.

■ *Example 6.1: Finding the DSA*

Suppose that the DSA is to be found from a centerpoint of (2,4) with a step size of 5, and the response surface were estimated by the following equation:

$$z = 13 + 8x - 3y + 2xy$$

Then the computation would proceed as follows:

Step	Action	Result
1	$R1 = b_1 + b_3 y_{cp}$	16
2	$R2 = b_2 + b_3 x_{cp}$	-6
3	$R3 = \sqrt{(R1)^2 + (R2)^2}$	17.09
4	$R4 = \text{Step} \dfrac{R1}{R3}$	4.68
5	$R5 = \text{Step} \dfrac{R2}{R3}$	-1.76
6	New $x_{cp} = x_{cp} + R4$	6.68
7	New $y_{cp} = y_{cp} + R5$	2.24

Lines 6 and 7 are the x and y coordinates of the centerpoint for the next experiment: $x = 6.68$, $y = 2.24$.

This worksheet was designed under the assumption that the predictors are in the same units; some scaling of predictors will be necessary in those cases where this assumption is false. This can often be accomplished simply by recording measurements in the unit appropriate to the predictor—recording critical dimensions (CD) in microns (μm), and gate oxide thickness in angstroms, for example. Units should be chosen so that a change of 1 unit for one predictor would be expected to have about the same order of magnitude of effect on the response as a change of one unit for the other predictor.

When this isn't true, some additional scaling of the predictors will be necessary. For example, if it really took a hundred-angstrom change in gate oxide thickness to produce a 1-V V_T shift, and this same effect would be caused by a 1-micron (1-μm) change in CD, then divide gate oxide thickness by 100 before undertaking any computations. Any scaling must done before the linear regression coefficients are produced. Once a new centerpoint is provided via the worksheet, it would be necessary to multiply gate oxide thickness by 100 to arrive at the original unit of measurement.

A factorial model that fits well on part of a response surface may fail to fit on another part of the same surface. Applying the factorial model to an unsuitable surface will result in a slower ascent to the desired optimum than would application of a more accurate model, and it might even lead the experimenter away from the optimum.

There are two easy ways to detect lack of fit: residual analysis, and the pure error lack of fit test. These will be applied within the examples which follow.

When lack of fit is discovered, a more suitable model must be found. The usual next step is to try a quadratic model based on data from the central composite design.

6.3.2   The Central Composite Design

The central composite design is a factorial augmented by centerpoints and axial points as shown in Figure 6.15. Neither the distance of the axial points from the centerpoint nor the number of centerpoints is arbitrary; they affect important attributes of coefficient estimates and predicted values. Table 6.1 gives some central composite designs with up to five predictors.

The central composite design with two predictors (x and y) and a single response (z) can estimate a model of the form:

$$z_i = \beta_0 + \beta_1 x_i + \beta_2 y_i + \beta_3 x_i y_i + \beta_4 x_i^2 + \beta_5 y_i^2 + \varepsilon_i$$

The mathematics underlying the DSA method is the same for central composite designs as it was for factorial designs: move in the direction of the gradient. Because the model is a little more complicated now, the gradient has a slightly different form:

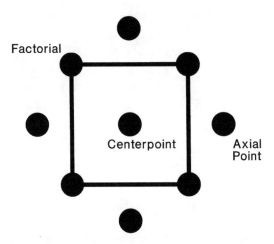

Fig. 6.15   This two-predictor central composite design is a 2^2 factorial augmented by centerpoints and axial points. The distance of the axial points from the center of the experiment, and the number of centerpoints are both important design criteria.

TABLE 6.1   Central Composite Design Choices*.

Predictors	Factorial Points	Axial Points	Axial Distance	Center Points	Total Design Points
2	4	4	1.414	5	13
3	8	6	1.682	6	20
4	16	8	2.000	7	31
5	16†	10	2.000	6	32

*Designs for several central composite experiments are given. The number of predictors determines the choice of the design: a three-factor design has eight factorial points, six axial points that are 1.682 steps from the centerpoint, and six centerpoints. The particular designs shown here are *uniform precision* designs. This means that predicted values will have the same variance at the centerpoint of the design as they will at a distance of one unit from the center.
†The-five predictor design utilizes a 2^{5-1} fractional factorial rather than a full 2^5 factorial.

$$\nabla (x, y) = \begin{pmatrix} b_1 + b_3\, y + 2\, b_4\, x \\ b_2 + b_3\, x + 2\, b_5\, y \end{pmatrix}$$

The computation is outlined in the worksheet below. The same caution about scaling applies for this worksheet as for the previous one: a change of one unit for either predictor should result in about the same magnitude of change in the response as an equivalent change in the other predictor.

Step	Action	Result
1	$R1 = b_1 + b_3 y_{cp} + 2b_4 x_{cp}$	R1
2	$R2 = b_2 + b_3 x_{cp} + 2b_5 y_{cp}$	R2
3	$R3 = \sqrt{(R1)^2 + (R2)^2}$	R3
4	$R4 = \text{Step} \dfrac{R1}{R3}$	R4
5	$R5 = \text{Step} \dfrac{R2}{R3}$	R5
6	New $x_{cp} = x_{cp} + R4$	new x_{cp}
7	New $y_{cp} = y_{cp} + R5$	new y_{cp}

■ *Example 6.2: The DSA for a Quadratic Model*

An experiment centered at $x = 1$, $y = -1$ produced data resulting in an estimated response equation of:

$$z = 12 + 6x + 4y - 2xy + 0.2x^2 - 0.3y^2$$

If step size were chosen to be 1.8, the next centerpoint would be at $x = 2.72$ and $y = 0.468$.

Step	Action	Result
1	$R1 = b_1 + b_3 y_{cp} + 2 b_4 x_{cp}$	8.4
2	$R2 = b_2 + b_3 x_{cp} + 2 b_5 y_{cp}$	2.6
3	$R3 = \sqrt{(R1)^2 + (R2)^2}$	8.79
4	$R4 = \text{Step } \dfrac{R1}{R3}$	1.72
5	$R5 = \text{Step } \dfrac{R2}{R3}$	0.532
6	New $x_{cp} = x_{cp} + R4$	2.72
7	New $y_{cp} = y_{cp} + R5$	−0.468

The central composite design differs from common two-level factorials in that it uses a quadratic model to estimate the effects of squared predictors, and thus better approximates curved surfaces. Most response surfaces in actual practice can be adequately modeled at least piecewise with the quadratic model.

There are situations where even a quadratic model will not fit an observed response surface, so it is important to be able to test the fit of the model. This is accomplished exactly as was done for the simpler factorial-based model using residual analysis or the pure error lack of fit test.

When the quadratic model fits poorly, consider the following options:

- If results continue to improve with each successive experiment, continue to use the same model.
- Try a smaller step size. This will often improve the apparent fit by allowing a smaller patch of the response surface to be approximated each time. There is considerable risk in this approach: the surface may in reality be so complicated that a quadratic model cannot find an optimum, or can only point to a local optimum.
- Examine residuals carefully. Outliers can often cause a Type I error (false positive) on a lack of fit test, even when the model fits well.
- Use some prior knowledge of the physical situation to choose a more appropriate model.
- Switch to an optimization method which does not require any model; the simplex method is one such method. Be prepared to pay a penalty in the number of iterations required when using the simplex method.

6.3.3   The Process of DSA

The DSA method is sequential, with all but the first experiment starting at a point suggested by a previous experiment. A checklist for the DSA process is shown in Figure 6.16, and a discussion of each point follows:

<u>The Initial Experiment</u>
☐   Choose the predictors.
☐   Choose a centerpoint which is likely to produce useful
    information.
☐   Choose a step size for each predictor.
☐   Consider all important responses.
<u>Assess the Experiment</u>
☐   Perform a regression analysis to obtain coefficients and
    ascertain statistical significance.
☐   Check model assumptions.
☐   Produce contour plots based on estimated regression
    equations.
<u>Plan the Next Experiment</u>
☐   Determine if another experiment is necessary.
☐   Determine the direction of steepest ascent for each
    response.
☐   The design for next experiment will be based on the shape of
    the response surface, as it is understood from data
    collected in prior experiments.
<u>Verify the Chosen Optimum</u>
☐   Assess process performance in the area around the optimum by
    running an experiment centered at that point.
☐   Run a production test.

Fig. 6.16   This checklist for the direction of steepest ascent method of optimization is used to plan a sequence of experiments in which knowledge from each experiment is used to plan the next experiment.

1. The Initial Experiment
    a. Choose the predictors.
    By this stage of experimentation, sufficient knowledge should have been obtained to fix settings for all qualitative factors, and continue optimization with only a few very important quantitative predictors.
    b. Choose a centerpoint that is likely to produce useful information.
    This may come from earlier experiments, or it may be determined by settings recommended by the equipment manufacturer. The centerpoint should not be at the limit of any allowable predictor setting, since design points will be chosen in all directions around the center.
    c. Choose a step size for each predictor.
    Step size should be large enough to have a noticeable effect on some response, but not so large that corners of the experiment might fall off a cliff. Step size can be changed in later experiments without any sacrifice of experimental efficiency
    d. Consider all important responses.
    Responses will often have competing optima, so it is important to make decisions about operating conditions with full knowledge of their individual behavior.

2. Assess the Experiment
    a. Perform a regression analysis to obtain coefficients and test their statistical significance.
    Because many experiments will be performed with small samples, predictors may seem statistically insignificant even when they are predicting quite well. The significance test is used more as an indicator than a true test in this case. The ulti-

mate test is whether the response continues to improve with each experiment. If it does, then the optimization process is probably working.

b. Check model assumptions.

A sufficiently large violation of model assumptions can produce misleading results. Check for curvature when using purely linear (factorial) designs, and if lack of fit is evident, make some additional runs to refine the model.

c. Produce contour plots based on estimated regression equations.

These plots (one for each response) help to check that the model makes good physical sense, and to choose the centerpoint for the next experiment.

3. Plan the Next Experiment

a. Determine if another experiment is necessary.

If process goals are satisfied for each response, or no further improvement seems possible, further experimentation is unnecessary. In this case, go on to item 4, Verify the Chosen Optimum.

b. Determine the DSA for each response.

If these vectors point in roughly the same direction, it is easy to choose the direction in which to move the centerpoint. If they point in differing directions, some compromises will have to be made. See Section 6.3.4 for some methods of reconciling competing responses.

c. The design for next experiment will be based on the shape of the response surface, as it is understood from data collected in prior experiments.

If the surface is believed to have important curvature, then a design like the central composite will be used to model the surface.

4. Verify the Chosen Optimum

a. Assess process performance in the area around the optimum by running an experiment centered at that point.

Optimum settings are not always those that produce the best response at the centerpoint of an experiment; the robustness of the optimum about the proposed target must be considered as well. Factorial or fractional factorial designs are well suited to assess this important quality.

b. Run a production test.

By running the optimum process in parallel with the standard process, a very discriminating comparison of the two can be made. Splitting production lots (see Chapter 3) is an especially effective type of experiment for this purpose. During this time, enough product should be manufactured to discover any unsuspected side effects.

The DSA method flow will be demonstrated in the context of a rhenium silicide process using a combination of factorial and central composite experiments. In the example which follows, a single response is optimized.

■ *Example 6.3: Optimizing a Single Response*

The first rhenium silicide deposition experiment was an unreplicated 2^2 factorial with five centerpoints. The center settings were chosen to be the present production settings of 90°C and 21.25 SCCM. The predictors were varied 2 units in each direction. Data from this first experiment is shown below:

Rhenium Silicide Deposition

The purpose of the rhenium silicide reactor is to deposit a thin layer of highly conductive rhenium silicide on the wafer surface. Rhenium silicide is produced by reacting silane and rhenium hexaflouride under conditions of carefully controlled temperature and pressure:

$$ReF_6 + 2SiH_4 \rightarrow ReSi_2 + 6HF + H_2$$

The process has been studied with screening experiments and some initial factorials, and at this time there are only two predictors of interest: ReF_6 flow [measured in standard cubic centimeters per minute (SCCM)], and chamber temperature.

The most important attribute of the deposited film is its resistivity, measured in ohms per square-no unit for the area is given because the measurement is always made on the area of fixed size and geometry. The ideal resistivity is 33, although anything from 24 to 51 is acceptable. Of nearly equal importance is the etch rate of the film, because it affects the selectivity of the following plasma etch. The ideal etch rate is 66 Å/s, with specified limits from 52 to 81 Å/s.

Deposition rate is an important economic parameter: the higher this rate, the faster the machine throughput.

One other critical response is the occurrence of the defect known as "green rocks," which are in fact crystals of ReF_2 sometimes produced by an unintentional side reaction. In addition to their undesirable visual appearance, green rocks can sometimes cause holes in the film because they etch much faster than $ReSi_2$. Green rocks can only be detected with a visual inspection immediately after deposition, and although they may at times be very numerous, operators only count up to 250 rocks per wafer.

The reaction chamber becomes coated with a polymer which is a side-product of the reaction, and this must be removed periodically with a ClF plasma clean.

TEMP	FLOW	RHO
86	17.25	51.90
86	25.25	64.95
90	21.25	51.50
90	21.25	49.35
90	21.25	50.85
90	21.25	49.05
94	17.25	42.95
94	25.25	51.55

The present state of knowledge about the two predictors and the responses can be summarized as follows:

- Increasing temperature or increasing flow will decrease resistivity.
- Decreasing flow will decrease etch rate, but there is some interaction with temperature.
- Increasing temperature will increase deposition rate.
- Not much is known about green rocks except that they seem to only occur at ReF_6 flows greater than 20 SCCM.

Response	*Resistivity*	*Etch Rate*	*Deposition Rate*	*Green Rocks*
Needs:	LSL = 24	LSL = 52	Bigger is better	Zero
	TGT = 33	TGT = 66		
	USL = 51	USL = 81		
Influences		*Correlation*		
Flow	Negative	Positive	?	Positive if flow > 20 SCCM
Temperature	Negative	?	Positive	?

A regression analysis of this data was performed assuming a linear model with interactions:

$$RHO_i = \beta_0 + \beta_1\,TEMP_i + \beta_2 FLOW_i + \beta_3 FLOW_i TEMP_i + \varepsilon_i$$

Regression coefficients were estimated by SAS, and are shown below with an analysis of variance:

Analysis of Variance

Source	df	Sum of Squares	Mean Square	F Value	Prb > F
Model	3	247.01187	82.33729	18.094	0.0086
Error	4	18.20187	4.55047		
C Total	7	265.21375			
Root MSE		2.13318	R-square		0.9314

Parameter Estimates

Variable	df	Parameter Estimate	Standard Error	Pr > |T|
INTERCEP	1	15.498828	129.85588019	0.9107
TEMP	1	0.080664	1.44144450	0.9581
LOW	1	7.610937	6.00549914	0.2738
TEM_FLO	1	−0.069531	0.06666196	0.3558

Abbreviations: df, degrees of freedom; MSE, mean square due to error.

The regression seems effective: 93% of the variance in the data is accounted for by the model.

The plot of residuals by temperature shown in Figure 6.17 indicates that the simple linear model used here does not fit exactly. Technically, this means that a different model should be fit to this data, but the factorial experiment really does not allow anything more complex than the one which has already been tried.

The experimenter here had two options: gather data suitable for quadratic modeling before planning where to center the next experiment, or use the present model and hope it fits well enough to point in a direction that will improve the process. Because the predictive ability of the model seems very good in spite of the lack of fit, the second option was chosen. However, since a quadratic model is probably needed, any later experiments will use the central composite design.

The DSA was computed using the worksheet to arrive at a centerpoint for the next experiment:

Step	Action	Result
1	$R1 = b_1 + b_3 y_{cp} + 2 b_4 x_{cp}$	−1.39687
2	$R2 = b_2 + b_3 x_{cp} + 2 b_5 y_{cp}$	1.35147
3	$R3 = \sqrt{(R1)^2 + (R2)^2}$	1.94480
4	$R4 = \text{Step} \dfrac{R1}{R3}$	−2.873
5	$R5 = \text{Step} \dfrac{R2}{R3}$	2.780
6	New $x_{cp} = x_{cp} + R4$	92.87
7	New $y_{cp} = y_{cp} + R5$	18.47

Resistivity was above target at the last centerpoint, so the next experiment will be centered in the opposite direction from the DSA. This is done by subtracting rather than adding in steps 6 and 7 of the worksheet. Allowing a vector length of 4 units, the centerpoint for the next experiment is determined to be:

$$\text{TEMP} = 93$$

$$\text{FLOW} = 18.5$$

To check the DSA computation and confirm that the predicted response surface has a shape consistent with previous scientific knowledge, a contour plot (Figure 6.18) was produced. Some care must be taken in interpreting such a plot because extrapolations outside the range of the original data may be nonsense.

The next experiment in this optimization series was a central composite design. With two predictors, the design requires four factorial points and four axial points.

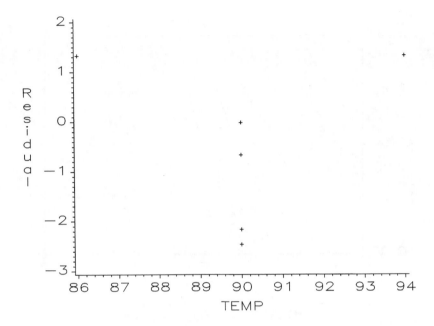

Fig. 6.17   This plot of resistivity residuals by temperature shows curvature. This probably means that quadratic terms were wrongly left out of the model.

Five centerpoints were run as recommended in Table 6.1. This experiment was run assuming that a quadratic model of the response surface is appropriate:

$$RHO_i = \beta_0 + \beta_1 \, TEMP_i + \beta_2 \, FLOW_i + \beta_3 \, TEMP_i \, FLOW_i + \beta_4 \, TEMP_i^2 + \beta_5 \, FLOW_i^2 + \varepsilon_i$$

The design is shown below; note that the axial points are a distance of 1.414 steps (square root of 2) from the center of the design.

	Temp	*Flow*
Factorial points	89.0	14.5
	97.0	14.5
	89.0	22.5
	97.0	22.5
Axial points	87.5	18.5
	98.5	18.5
	93.0	13.0
	93.0	24.0
Center points	93.0	18.5
	93.0	18.5
	93.0	18.5
	93.0	18.5
	93.0	18.5

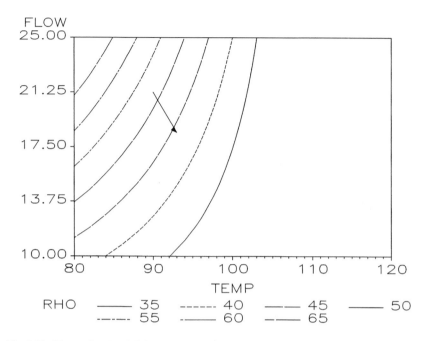

Fig. 6.18   The predicted resistivity response surface is shown with temperature on the *x*-axis and flow on the *y*-axis. The DSA vector is marked with an arrow originating at the center of the experiment.

Raw data from the experiment is shown below:

TEMP	FLOW	RHO
87.5	18.5	52.60
89.0	14.5	50.25
89.0	22.5	55.90
93.0	13.0	46.40
93.0	18.5	46.05
93.0	18.5	46.50
93.0	18.5	47.10
93.0	18.5	47.75
93.0	18.5	48.65
93.0	24.0	54.05
97.0	14.5	43.20
97.0	22.5	46.20
98.5	18.5	41.85

A regression analysis based on the following quadratic model was used:

$$\text{RHO}_i = \beta_0 + \beta_1\,\text{TEMP}_i + \beta_2\,\text{FLOW}_i + \beta_3\,\text{TEMP}_i\,\text{FLOW}_i + \beta_4\,\text{TEMP}_i^2 + \beta_5\,\text{FLOW}_i^2 + \varepsilon_i$$

A summary of the results of a regression analysis based on this model is shown below:

Regression	Degrees of Freedom	Type I Sum of Squares	R-Square	Pr > F
Linear	2	174.992540	0.8804	0.0000
Quadratic	2	16.882292	0.0849	0.0061
Cross product	1	1.755625	0.0088	0.1657
Total regress	5	193.630457	0.9742	0.0000

The model seems to have good predictive ability because 97% of the variance in resistivity is accounted for by terms in the model. The quadratic terms play an important part in this prediction (*P*-value of 0.0061), but the crossproduct terms are not as important (*P*-value of 0.1657).

Because there were concerns in the previous experiment about lack of fit, a pure error lack of fit test was performed; results are shown below:

Residual	Degrees of Freedom	Sum of Squares	Mean Square	Prb > F
Lack of fit	3	0.906774	0.302258	0.8341
Pure error	4	4.227000	1.056750	
Total error	7	5.133774	0.733396	

Lack of fit seems to be insignificant now.

The regression coefficients obtained from this analysis are shown below:

Parameter	Degrees of Freedom	Parameter Estimate
INTERCEPT	1	108.768336
TEMP	1	−0.578788
LOW	1	0.731230
TEMP*TEMP	1	0.001784
FLOW*TEMP	1	−0.041406
FLOW*FLOW	1	0.100957

These results are used in the worksheet to determine the direction of steepest ascent and choose a centerpoint for the next experiment:

Step	Action	Result
1	$R1 = b_1 + b_3 y_{cp} + 2b_4 x_{cp}$	−1.0130
2	$R2 = b_2 + b_3 x_{cp} + 2b_5 y_{cp}$	0.6159
3	$R3 = \sqrt{(R1)^2 + (R2)^2}$	1.1855
4	$R4 = \text{Step}\ \dfrac{R1}{R3}$	−3.4179
5	$R5 = \text{Step}\ \dfrac{R2}{R3}$	2.0780
6	New $x_{cp} = x_{cp} + R4$	96.42
7	New $y_{cp} = y_{cp} + R5$	16.42

Figure 6.19 is a contour plot of the predicted response surface annotated with the DSA vector pointing to the centerpoint of the next experiment.

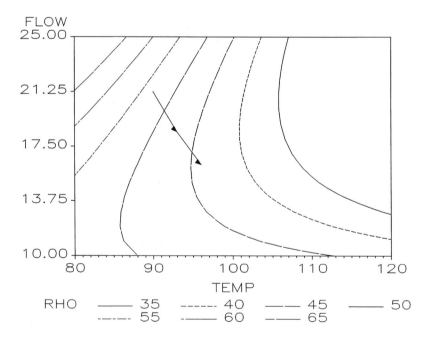

Fig. 6.19   The predicted resistivity response surface is shown with temperature on the *x*-axis and flow on the *y*-axis. The DSA vectors from the first two experiments are marked with arrows.

The process used in the example above is typical of optimization by DSA: simple factorial experiments are used until they fail to fit, and then central composite experiments are employed. The process continues until optimization goals are achieved, or no further improvement is observed.

6.3.4   Multiple Response Problems

A process rarely has just one important response, so any practical optimization method has to be able to account for multiple responses. Each response may have different optimum process settings, so some compromises and tradeoffs will have to be made to arrive at settings producing satisfactory results for each individual response.

One approach to resolving conflicts of this sort is to optimize some summary *figure of merit* (FOM) instead of trying to optimize individual responses. For example, if both a high etch rate and a low throughput time were desired from the rhenium silicide reactor, then one figure of merit might be etch rate minus throughput time. Optimizing a FOM should result in a process that is acceptable with respect to each individual response, and economically most beneficial to the process as a whole.

A difficulty with the FOM approach is in finding a summary measure that meets these criteria. Too often, a FOM that seems reasonable when formulated will result in a false optimum—either because some important response is not within a range of acceptability, or because a more cost-effective solution with nearly equivalent individual response values could be obtained. Optimizing *only* a FOM merit has the additional disadvantage that engineering judgment cannot be applied to the responses individually.

A more effective strategy is to carefully consider the DSA vectors suggested by each response, and then to somehow arrive at a direction that will improve some important responses without significantly degrading others. A FOM may still be useful when no direction is obvious, or when the situation is simply too complicated to look at each response individually.

This approach will be exercised in the example below on the rhenium silicide process with respect to its four important responses.

■ *Example 6.4: Optimizing Multiple Responses*

Two of those rhenium silicide process responses have a target, and a region about the target in which operation would be acceptable. There is a response which must be nearly zero—the number of green rocks. There is also a "bigger is better" response—deposition rate.

The design chosen for the first optimization experiment was a two-factor central composite design. Results from the experiment are shown below:

TEMP	FLOW	RHO	ERATE	DEPRATE	ROCKS
87.5	18.5	52.60	80.9	80.5	15
89.0	22.5	56.60	96.8	87.9	250
89.0	22.5	55.90	96.1	85.1	250
93.0	13.0	46.40	51.1	83.5	250
93.0	18.5	46.05	69.9	83.4	0
93.0	18.5	46.50	68.7	84.2	5
93.0	18.5	47.10	70.4	88.0	0
93.0	18.5	47.75	69.7	87.2	4
93.0	18.5	48.65	68.4	86.3	0
93.0	24.0	54.05	100.3	94.5	0
97.0	14.5	44.95	47.3	83.8	0
97.0	14.5	43.20	47.3	81.9	0
98.5	18.5	41.85	63.7	87.6	26

A regression analysis was done separately for each response. The results for resistivity show that both linear and quadratic effects are important:

Analysis of Variance for Variable RHO

Regresion	Degrees of Freedom	Type I Sum of Squares	R-Square	Pr > F
Linear	2	234.805683	0.8704	0.0000
Quadratic	2	25.050403	0.0929	0.0039
Cross product	1	3.435145	0.0127	0.0952
Total regression	5	263.291231	0.9760	0.0000

Parameter	Degrees of Freedom	Parameter Estimate
INTERCEPT	1	14.663717
TEMP	1	0.542484
Flow	1	4.802503
TEMP*TEMP	1	0.000496
FLOW*TEMP	1	−0.084366
FLOW*FLOW	1	0.099669

For etch rate, linear, crossproduct, and quadratic effects are all significant:

Analysis of Variance for Variable ERATE				

Regression	Degrees of Freedom	Type I Sum of Squares	R-Square	Pr > F
Linear	2	3773.725623	0.9838	0.0000
Quadratic	2	47.893019	0.0125	0.0001
Cross product	1	10.768404	0.0028	0.0020
Total regression	5	3832.387046	0.9991	0.0000

Parameter	Degrees of Freedom	Parameter Estimate
INTERCEPT	1	1286.298463
TEMP	1	−22.071916
FLOW	1	−17.063770
TEMP*TEMP	1	0.095207
FLOW*TEMP	1	0.149372
FLOW*FLOW	0.207603	

The analysis of deposition rate reveals significant effects for linear terms alone:

Analysis of Variance for Variable DEPRATE				

Regression	Degrees of Freedom	Type I Sum of Squares	R-Square	Pr > F
Linear	2	98.815027	0.6517	0.0023
Quadratic	2	24.756500	0.1633	0.0673
Cross product	1	6.739923	0.0445	0.1803
Total regress	5	130.311450	0.8595	0.0068

Parameter	Degrees of Freedom	Parameter Estimate
INTERCEPT	1	−256.911268
TEMP	1	9.308009
LOW	1	−13.845244
TEMP*TEMP	1	−0.058512
FLOW*TEMP	1	0.118174
FLOW*FLOW	1	0.105124

The analysis for rocks did not find any statistically significant effects, but this is not really surprising: green rocks act more like a categorical response here than a continuous one (they are present, or they are not), so the ANOVA assumptions are not satisfied. The ANOVA is shown below, along with a lack of fit test indicating that the model for rocks is not appropriate.

Analysis of Variance for Variable ROCKS

Regression	Degrees of Freedom	Type I Sum of Squares	R-Square	Pr > F
Linear	2	31,725	0.2279	0.2359
Quadratic	2	41,006	0.2946	0.1696
Cross product	1	4,369.997020	0.0314	0.5055
Total regress	5	77,101	0.5538	0.2442

Residual	Degrees of Freedom	Sum of Squares	Mean Square	Pr > F
Lack of Fit	1	62,086	62,086	0.0000
Pure Error	6	24.800000	4.133333	
Total error	7	62110	8,872.903979	

Parameter	Degrees of Freedom	Parameter Estimate
INTERCEPT	1	3283.277851
TEMP	1	−76.979660
FLOW	1	125.093296
TEMP*TEMP	1	0.618182
FLOW*TEMP	1	−3.009091
FLOW*FLOW	1	4.072727

The direction of steepest ascent was computed as usual for each response: these directions are shown on contour plots for etch rate (Figure 6.20), deposition rate (Figure 6.21), and green rocks (Figure 6.22).

The response surface contour plot for resistivity shown in Figure 6.23 shows the DSA for resistivity along with those for the other three responses. There is a conflict in the directions that would improve each response. Two approaches to resolving this conflict are demonstrated below.

Considering responses one at a time:

- Resistivity could be improved, but it is within specifications. Any change that would not move resistivity farther from target would probably be acceptable.
- It appears that etch rate would be moved past its target if a full step were taken in the etch rate DSA. Moving a full step in a direction of increasing temperature (so

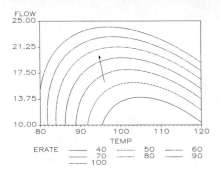

Fig. 6.20   The DSA vector for the etch rate (ERATE) response is shown here.

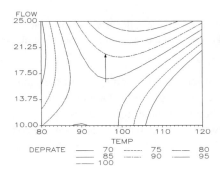

Fig. 6.21   The DSA vector for the deposition rate (DEPRATE) is shown in this contour plot.

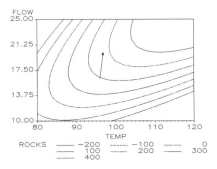

Fig. 6.22   The DSA vector for green rocks is shown in this contour plot. Because this model exhib-ited extreme lack of fit, neither the plot nor the DSA vector should be applied without independent information confirming that this is a desirable direction.

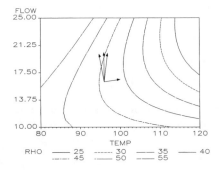

Fig. 6.23   The response surface contour plot for resistivity is shown here with the DSA for resistiv-ity (arrow pointing to the right) along with directions for the other three responses. No single direc-tion will improve all three responses most rapidly.

as not to degrade resistivity) and increasing flow to put etch rate exactly on target would result in settings of:

$$\text{Temperature} = 99$$
$$\text{Flow} = 18.75$$

The anticipated effect of this change is shown in Figure 6.24.

The remaining responses will now be checked to see if these suggested settings are reasonable.

- Deposition rate will increase at the new settings, which is desirable.
- Rocks should decrease at the new settings, and this is also an improvement in the process.

The FOM approach requires that some overall measure of goodness be computed on the basis of all four responses. One FOM that seems reasonable here is based on the cost of being off target for each response. If each individual cost is formulated so it is always between 0 and 1, then one choice for an aggregate measure is four minus the sum of the costs. This FOM will take values between 0 (worst) and 4 (best):

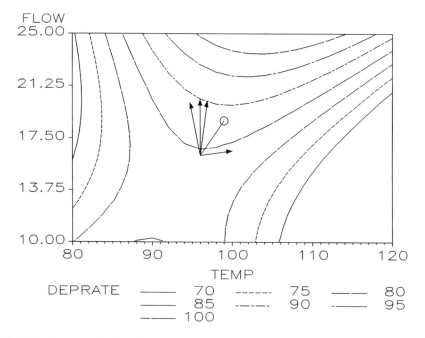

Fig. 6.24   A compromise direction of improvement is denoted by the vector ending in a circle on this contour plot for deposition rate.

$$FOM = 4 - Cost_{RHO} - Cost_{DEPRATE} - Cost_{ETCHRATE} - Cost_{Rocks}$$

The loss function chosen for resistivity is the upside-down normal loss function (UDNLF) assuming 50% loss at the lower specification limit. Details on this type of loss function can be found in Drain and Gough (1996).

$$Loss\ (RHO) = 1 - e - \frac{(RHO - 33)^2}{117}$$

See Figure 6.25 for a graph of this loss function.
The loss function for etch rate is defined similarly:

$$Loss\ (rate\ ) = 1 - e - \frac{(rate - 66)^2}{70.8}$$

The loss function for deposition rate is linear:

$$Loss\ (dep) = 1 - \frac{dep}{100}$$

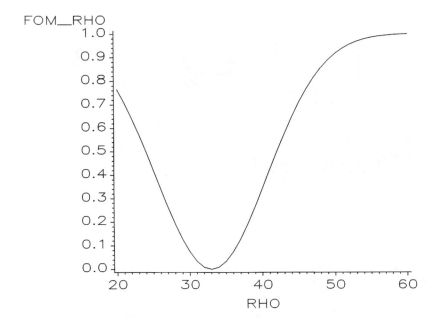

Fig. 6.25   The upside-down normal loss function was chosen for resistivity; 50% loss at the lower specification limit is assumed.

The loss function for green rocks is chosen to have a large penalty for even a few rocks:

$$\text{Loss (rocks)} = \sqrt{\frac{\text{rocks}}{250}}$$

A linear regression with the FOM as its only response (not shown) suggests the DSA annotated with a circle on the contour plot of Figure 6.26. This is different from the direction that was suggested by a careful consideration of each response individually.

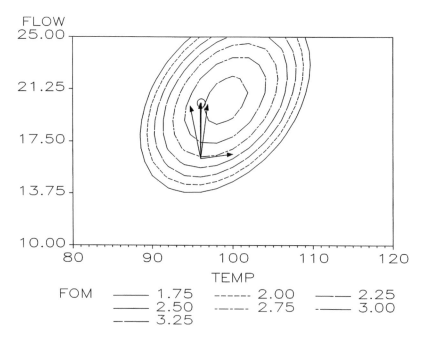

Fig. 6.26   This contour plot of the figure of merit suggests the DSA annotated with a circle. This differs from the direction suggested by a careful consideration of each response individually.

6.3.5   DSA in Higher Dimensions

DSA optimization with more than two predictors is theoretically exactly the same as that with just two predictors, but practical considerations require small adjustments in the application of the theory.

Because it is difficult for most people to visualize surfaces in more than three dimensions, contour plots are less useful when more than two predictors are used. DSA requires a heavier reliance on mathematics than before, and the extra dimensions make the application of engineering judgment difficult.

With more than two predictors, factorial experiments and the central composite experiments built from them may become unwieldy. Fractional factorials are used wherever possible to reduce the size of experiments, but those experiments must still have sufficient resolution to estimate important interactions.

The direction of steepest ascent method with five predictors will be demonstrated in the example which follows:

■ *Example 6.5: The DSA in Higher Dimensions*

Suppose that three additional predictors were considered in the rhenium silicide reaction optimization. These parameters are thought to affect the ClF plasma clean of the chamber between reactions:

- Pressure during the clean
- ClF flow during the clean
- Power during the clean

For the first optimization experiment the centerpoint is chosen to be:

$$\text{Temperature} = 96°C$$
$$\text{ReF}_6 \text{ flow} = 16.25 \text{ SCCM}$$
$$\text{Pressure} = 24 \text{ mtorr}$$
$$\text{ClF flow} = 32 \text{ SCCM}$$
$$\text{Power} = 40 \text{ W}$$

A central composite design using a 2^{5-1} fractional factorial and six centerpoints will be performed. Axial points are two steps from the centerpoint. This design requires 10 axial points, so the total experiment requires 32 runs. Pressure was scaled so that one step is 3 mtorr; ClF was scaled so that one step is 4 SCCM; Power was scaled so that one step is 4 W.

Raw data from the experiment is shown below:

TEMP	FLOW	PRESS	ClF	POWER	RHO
88	16.25	24	32	40	65.30
92	12.25	20	28	44	77.75
92	12.25	20	36	36	70.05
92	12.25	28	28	36	79.60
92	12.25	28	36	44	80.60
92	20.25	20	28	36	47.00
92	20.25	20	36	44	51.75
92	20.25	28	28	44	52.65
92	20.25	28	36	36	49.95
96	8.25	24	32	40	79.75
96	16.25	16	32	40	49.45
96	16.25	24	24	40	58.30
96	16.25	24	32	32	45.95
96	16.25	24	32	40	52.15
96	16.25	24	32	40	53.90
96	16.25	24	32	40	52.70
96	16.25	24	32	40	53.75
96	16.25	24	32	40	53.80
96	16.25	24	32	40	50.95
96	16.25	24	32	48	60.70
96	16.25	24	40	40	52.20
96	16.25	32	32	40	55.60
96	24.25	24	32	40	52.15
100	12.25	20	28	36	64.45
100	12.25	20	36	44	69.15
100	12.25	28	28	44	77.15
100	12.25	28	36	36	61.90
100	20.25	20	28	44	39.05
100	20.25	20	36	36	40.80
100	20.25	28	28	36	39.60
100	20.25	28	36	44	44.80
104	16.25	24	32	40	39.90

These results of a linear regression analysis of this data will be the basis for determining the direction of steepest ascent:

Regression	Degrees of Freedom	Type I Sum of Squares	R-Square	Pr > F
Linear	5	3944.508854	0.8302	0.0000
Quadratic	5	420.446506	0.0885	0.0511
Cross product	10	95.676563	0.0201	0.9398
Total regress	20	4460.631922	0.9388	0.0004

Parameter	Degrees of Freedom	Parameter Estimate
INTERCEPT	1	665.642843
TEMP	1	−6.245204
LOW	1	−9.145253
PRESS	1	2.603254
ClF	1	−6.026147
POWER	1	−5.275176
TEMP*TEMP	1	0.023295
LOW*TEMP	1	−0.008008
LOW*FLOW	1	0.231889
PRESS*TEMP	1	−0.023242
PRESS*FLOW	1	−0.035742
PRESS*PRESS	1	0.022124
ClF*TEMP	1	0.002930
ClF*FLOW	1	0.101367
ClF*PRESS	1	−0.058398
ClF*ClF	1	0.064702
POWER*TEMP	1	0.027148
POWER*FLOW	1	−0.070508
POWER*PRESS	1	0.035352
POWER*ClF	1	0.028711
POWER*POWER	1	0.034624

From these regression coefficients the gradient vector is found to be:

$$\nabla f \,(\text{Temp, Flow, Press, ClF, Power}) = \begin{pmatrix} -1.07698 \\ -2.81201 \\ 0.398511 \\ 0.209837 \\ 0.722397 \end{pmatrix}$$

Since resistivity is still above target, the centerpoint for the next experiment is obtained by moving from the centerpoint against this vector 4 units (the step size):

$$
\begin{pmatrix} 96 \\ 16.25 \\ 24 \\ 32 \\ 40 \end{pmatrix} + \begin{pmatrix} 1.4 \\ 3.6 \\ -2.5 \\ 1.4 \\ -3.5 \end{pmatrix} = \begin{pmatrix} 97.4 \\ 19.85 \\ 21.5 \\ 33.4 \\ 36.5 \end{pmatrix}
$$

To help understand the effect such a change might have, contour plots of each of the 10 two-dimensional projections of the response space are shown in Figures 6.27–6.36. Note that CIF is denoted CLF in these figures. The DSA vector is marked in each plot.

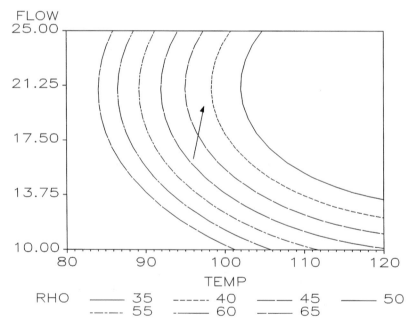

Fig. 6.27 Contour plot of resistivity by temperature and flow. The DSA is marked with an arrow that originates at the center of the experiment.

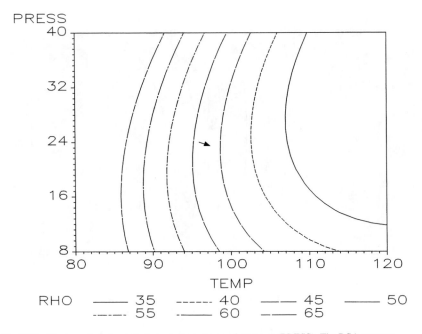

Fig. 6.28   Contour plot of resistivity by temperature and pressure (PRESS). The DSA appears smaller here because it is projected onto a plane where it changes little.

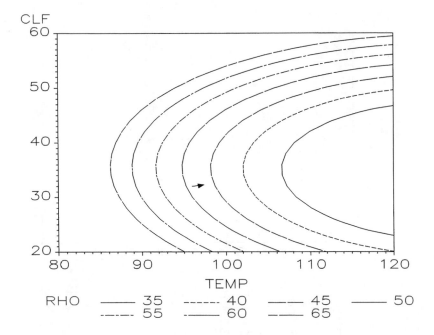

Fig. 6.29   Contour plot of resistivity by temperature and ClF.

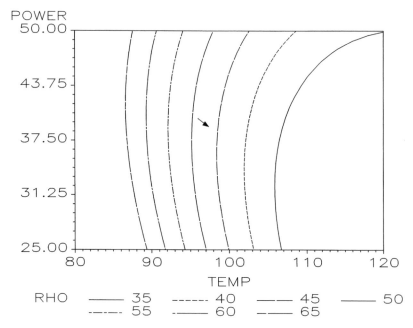

Fig. 6.30   Contour plot of resistivity by temperature and power.

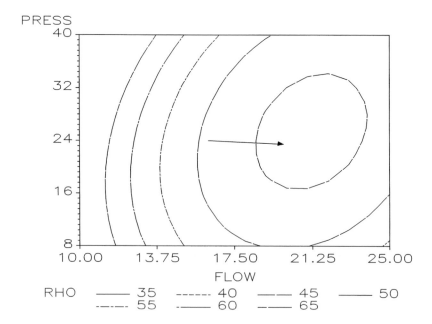

Fig. 6.31   Contour plot of resistivity by flow and pressure (PRESS).

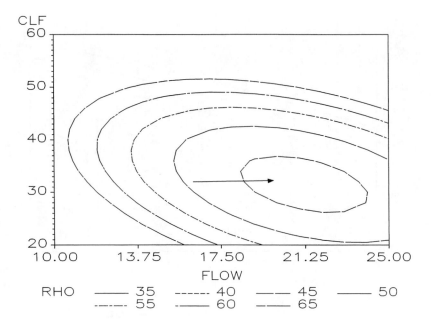

Fig. 6.32   Contour plot of resistivity by flow and ClF.

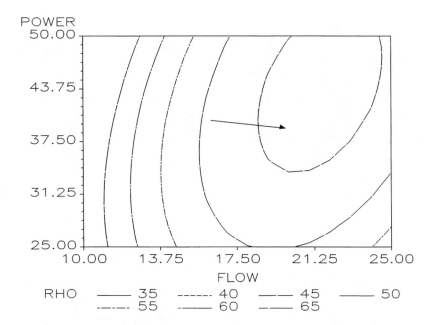

Fig. 6.33   Contour plot of resistivity by flow and power.

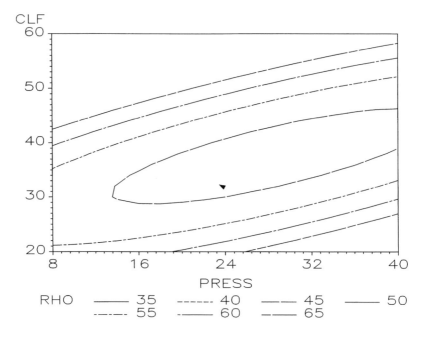

Fig. 6.34 Contour plot of resistivity by pressure and ClF. The direction of steepest ascent appears very small for this combination of predictors because it is almost perpendicular to the plane determined by them.

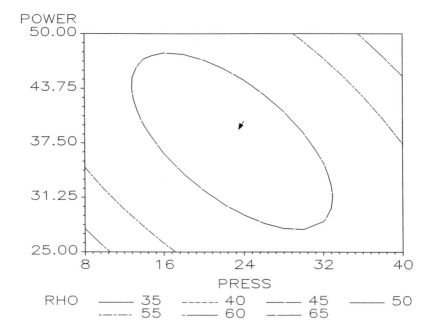

Fig. 6.35 Contour plot of resistivity by pressure and power.

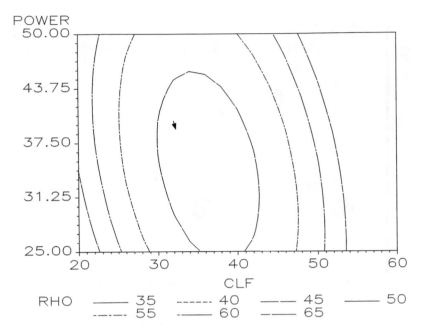

Fig. 6.36   Contour plot of resistivity by ClF and power. To help understand the effect such a change might have, contour plots of each of the 10 two-dimensional projections of the response space are shown in Figures 6.27–6.36. The DSA vector is marked in each plot

6.4   THE SIMPLEX METHOD

The simplex method is an entirely different approach to optimization which can be very successful under certain conditions. If settings are easy to change, if individual runs of the experiment can be executed quickly, and if random errors are relatively small the simplex method may be more efficient than the direction of steepest ascent method.

The simplex method takes its name from the geometrical term *simplex:* a regular figure with $n + 1$ vertices in n dimensions. The essence of the method can be illustrated for the two-predictor case with simple picture (Figure 6.37). In this two-predictor example, three combinations of predictor settings in a simplex arrangement are tried initially, and the response at each is noted. The next experiment requires only one run, and the settings for that run are determined by "reflecting" the experiment away from the setting combination with the worst response in the initial experiment (point W in Figure 6.38).

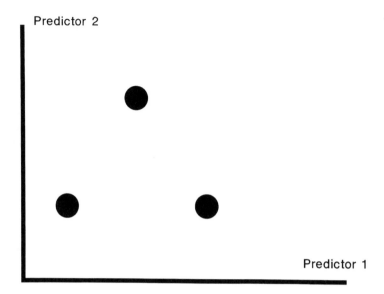

Fig. 6.37   The three points shown form a simplex in the plane determined by Predictor 1 and Predictor 2. If the points were connected, they would form an equilateral triangle.

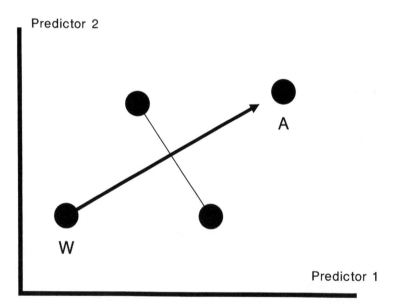

Fig. 6.38   The response was evaluated at each of the three points shown in Fig. 6–37, and the worst of the three points is marked with a "W." The predictor settings for the next experiment are determined by reflecting the experiment away from the setting combination with the worst response. This is equivalent to picking up the triangle at the W point, and pushing it as far as it could go away from the W point without distorting the simplex shape of the original experiment. In this case, the reflection away from the W point arrives at a point marked with an A.

The reflection away from point W produces point A, which is the only run required for the next experiment. The next experiment evaluates the response at this new point, reuses the two best points from the first experiment, and again reflects away from the worst of the three (point X in Figure 6.39) to get point B. The process is continued until a suitable optimum is obtained.

The generalization to more than two predictors is easy: if n predictors are used, then $n + 1$ runs are included in the initial experiment, and *only one additional run is needed for each following experiment.* The thriftiness of this method is immediately obvious: no matter how many predictors are included, each successive experiment requires only one additional run.

The simplex method is only effective when applied in very iteration-tolerant experimental situations: the time from the execution of an experiment to analysis of the results must be relatively short.

The method also requires that the response variance at any particular setting combination be rather small. Otherwise many false leads and unnecessary iteration will result.

The simplex method requires no knowledge of the actual functional relationship between predictors and the response. This is both a benefit and a detriment: extremely complex functional relationships can be ignored and good results still

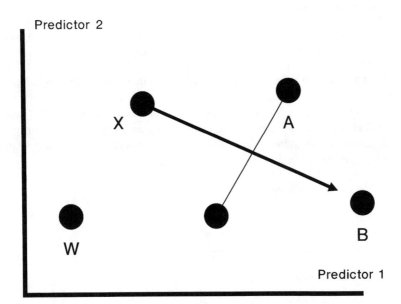

Fig. 6.39   The next experiment evaluates the response at the new point (A), reuses the two best points from the first experiment, and again reflects away from the worst of the three points (point X).

obtained with the method, but generalizing results away from the particular setting combinations already tried can be risky. Some other form of experiment—a fractional factorial perhaps—is necessary to assess the robustness of the optimum.

Since no arithmetic is performed on the responses, the method requires only that responses can be *ranked*—no actual numeric value is required. Thus the simplex method may be applied in cases unamenable to DSA—taste testing, or brand preference, for example.

The simplex method has its greatest advantage with respect to DSA when applied in cases with many predictors. No matter how many predictors are used, only one additional run is required for each experiment, and since the method does not rely on visualization, additional predictors are no impediment to its application.

The simplex method has been applied in:

- Optimizing the use of measurement equipment. Responses could be the time required for the measurement, repeatability, or accuracy. Since most modern measurement equipment has many user-adjustable settings, it can be difficult to apply the direction of steepest ascent method here.
- Optimizing the use of a computer system by tuning its operating system parameters. Each run of the experiment may take only minutes, settings are easy to change, and most computer systems can automatically monitor their own performance. The entire optimization exercise could itself be automated and rerun periodically to compensate for changes in usage of the system.
- An etch process with five continuously variable settings could take advantage of simplex optimization: each test wafer takes less than two minutes to etch, and critical measurements can be made within a few minutes of the etch.

One of the greatest advantages of the simplex method is the simplicity of its application. After the initial experiment is done, a bit of simple arithmetic determines settings for the next run. The simplex method flow is shown in the checklist below. This checklist for simplex optimization is used to plan a sequence of experiments that require only one additional run after the initial experiment and that systematically improve the desired response.

1. The Initial Experiment
   a. Choose the predictors.
   b. Choose a centerpoint that is likely to produce useful information.
   c. Choose a step size for each predictor.
   d. Consider all important responses.
   e. Design the initial experiment.
   f. Run the initial experiment.

2. Assess the Experiment
   a. Evaluate the response at each of the experimental settings.
3. Plan the Next Experiment
   a. Determine if another experiment is necessary.

b. Check for oscillation.

c. "Reflect away" from the run with worst response of the last n + 1 runs.

d. Replace "worn-out" design points.

4. Verify the Chosen Optimum

   a. Assess process performance in the area around the optimum by running an experiment centered at that point.

   b. Run a production test.

This checklist will now be discussed in detail.

1. The Initial Experiment

   a. Choose the predictors.

   Although the simplex method does not pay the major penalty which the DSA does for each additional predictor, it will take more iterations for each predictor added. Use screening experiments or process knowledge to select only (continuously variable) predictors which are likely to have an effect on the response.

   b. Choose a centerpoint that is likely to produce useful information.

   This centerpoint may come from earlier experiments, or it may be a setting combination recommended by the equipment manufacturer. The centerpoint should not be at the limit of any allowable predictor setting, since design points will be chosen in all directions around the center. The nearer the centerpoint is to the optimum, the quicker (and more surely) the method will arrive at that optimum.

   c. Choose a step size for each predictor.

   Step size should be large enough to have a noticeable effect on the response, but not so large that corners of the experiment might fall off a cliff. Simplex experiments usually have a smaller step size than would DSA experiments with the same predictors because the method uses no assumptions about the functional relationship between predictors and responses. A step for any predictor should have roughly the same importance (effect on the response) as for any other, so some prescaling may be necessary.

   d. Consider all important responses.

   Often, several indicative responses and some economic responses will have different optima, so it is important to make decisions about operating conditions with full knowledge of their behavior.

   e. Design the initial experiment.

   This is the only nontrivial part of the simplex method, but it is not really difficult. The initial simplex design with n predictors is an n + 1 run design. So, a design with six predictors would have seven runs. Once the predictors, the centerpoint, and the step size are known, use Table 6.2 to select settings for each of the n + 1 points in the initial design.

   f. Run the initial experiment.

   Randomize the runs in this first experiment carefully, and be especially vigilant for confounding. Because following experiments have only one run each, confounding will be practically impossible to detect after the initial experiment.

2. Assess the Experiment.

TABLE 6.2. Predictor Settings for Simplex Experiments

Run						
	Predictor					
	1	*2*	*3*	*4*	*5*	*6*
1	−0.5000	−0.2887	−0.2041	−0.1581	−0.1291	−0.1091
2	0.5000	−0.2887	−0.2041	−0.1581	−0.1291	−0.1091
3	0.0000	0.7113	−0.2041	−0.1581	−0.1291	−0.1091
4	0.0000	0.0000	0.7959	−0.1581	−0.1291	−0.1091
5	0.0000	0.0000	0.0000	0.8419	−0.1291	−0.1091
6	0.0000	0.0000	0.0000	0.0000	0.8709	−0.1091
7	0.0000	0.0000	0.0000	0.0000	0.0000	0.8909

Simplex experiments with n predictors start off with an $n + 1$-run experiment. This table gives the predictor setting combinations for that initial experiment.

Suppose a two-predictor experiment is being planned, then the first two columns and the first three rows of the table give the predictor settings: (−0.5000, −0.2887), (0.5000, −0.2887), (0.0000, 0.7113). The initial number of runs is always one more than the number of predictors.

The first predictor will take a value half a step below the centerpoint in the first run, half a step above the centerpoint in the second run, and will be set to the centerpoint on the third run.

    a. Evaluate the response at each of the experimental settings.

        For all but the initial experiment, only an single run is made; the response needs to be evaluated for that run only.

  3. Plan the Next Experiment

    a. Determine if another experiment is necessary.

        If process goals are satisfied, or no added improvement seems possible, further experimentation is unnecessary. In this case, go on to verify the chosen optimum.

    b. Check for oscillation.

        In some circumstances, the simplex method will never arrive at an optimum— there might not be an optimum, the local variance may be so large as to make finding an optimum unlikely, or the step size might be too large to detect an optimum. A fractional factorial centered at the last simplex run may be necessary to clarify the situation.

    c. ''Reflect away'' from the run with worst response of the last n + 1 runs.

        A simple algorithm is used to reflect from the worst point, and it works for any number of predictors. Suppose that n predictors are included in the design, and one design point had the worse response—call this point the W-point. Reflection away from the W-point proceeds as follows:

    (1) Add up predictor values over the last n + 1 points excluding the W-point.

    (2) Divide this sum (for each predictor) by n/2.

    (3) Subtract the predictor settings of the W-point from this sum.

    d. Replace "worn out" design points.

        Because data from each run is used to determine the direction for following runs, one aberrant observation can seriously mislead the experimenter. To limit the damage caused by this sort of misdirection, do not reuse a design point for more than n times, where n is the number of predictors. If that setting combination is needed more than n times, rerun it.

4. Verify the Chosen Optimum
   a. Assess process performance in the area around the optimum by running an experiment centered at that point.
   The simplex method itself provides little proof of robustness about a chosen optimum: a fractional factorial can give much more assurance that the optimum is indeed acceptable.
   b. Run a production test.
   By running the "optimum" process in parallel with the standard process, a very critical comparison of the two can be made. Splitting production lots (see Chapter 3) is an especially effective type of experiment for this purpose. During this time, sufficient product should be manufactured to discover any unsuspected side effects.

■ *Example 6.6: Initial Simplex Design for a Three-Factor Experiment*

An experiment has three predictors: P1, P2, and P3, with centerpoints and step sizes as follows:

	Predictor		
	P1	*P2*	*P3*
Centerpoint	50	110	20
Step size	2	12	4

The numbers shown in Table 6.2 are the number of steps away from the center for each predictor. Each row of the table denotes a run of the experiment; each column corresponds to a predictor. Since this design has only three predictors, only the first three columns and the first four rows of the table will be used:

		Predictor	
Run	*P1*	*P2*	*P3*
1	−0.5000	−0.2887	−0.2041
2	0.5000	−0.2887	−0.2041
3	0.0000	0.7113	−0.2041
4	0.0000	0.0000	0.7959

For each predictor, add or subtract the number of steps in the table to the center value for that predictor. For P1, the first design point would take a value of 49 (centerpoint minus half a step). Using this same technique on the rest of the predictors and design points yields the following simplex:

	Predictor		
Run	1	2	3
1	49.00	106.54	19.184
2	51.00	106.54	19.184
3	50.00	118.54	19.184
4	50.00	110.00	23.184

■ *Example 6.7: Reflection from the Worst Setting Combination*

Suppose that the experiment in the previous experiment had been executed and responses obtained. To find the settings for the next run, first rewrite the simplex runs in the order of ascending response values:

	Predictor			
Run	1	2	3	Response
1	51.00	106.54	19.184	112
2	49.00	106.54	19.184	115
3	50.00	118.54	19.184	119
4	50.00	110.00	23.184	122
5	48.33	116.85	21.851	

Add setting values for all but the worst run (which is in the first row of the table), divide by 3/2, and then subtract the settings for the worst run. For predictor 1 the computation looks like this:

$$\frac{49.00 + 50.00 + 50.00}{\frac{3}{2}} - 51.00 = 48.33$$

There are some situations in which reflection away from the design point with the worst response is not the best choice:

• The reflection would take some predictor setting past allowable or possible bounds. In this case, try reflection away from the *second worst* point.
• The reflection would take the experiment back to a setting that was rejected in the previous experiment. Again, try reflection away from the second worst point.

The simplex method will be demonstrated in the example below, which uses six predictors, and has a FOM as its response.

"The Fast Inspection Machine"

The Fast Inspection Machine (FIM) is an intelligent optical comparator which can quickly locate visual defects less than 1 micron (1 μm) in diameter. Images of those defects are saved for later classification (by a human operator) and statistical analysis.

Inspection parameters must be adjusted for each type of wafer to be examined so that defects of interest are detected often, and nuisance or false defects are rarely detected. By making sure that only important defects are seen, classification time and inspection time can both be decreased. Six such parameters can be adjusted by the machine programmer:

1. Optical parameters affect the collection of defect information.
   a. Camera contrast is adjustable from 0 to 255.
   b. Brightness of the inspection camera is adjustable from 0 to 255.
2. Parameters used in the FIM analysis algorithm to make decisions about potential defects. Each of these parameters is adjustable from 0 to 100.
   a. P12 adjusts sensitivity for straight edges, as seen in Fig. 6–40
   b. P34 adjusts for corners, as seen in Fig. 6–41.
   c. P56 adjusts for voids, as seen in Fig. 6–42.
   d. P78 adjusts for protrusions, as seen in Fig. 6–43.

   Present knowledge of the relationship between defect characteristics, parameter settings, and probability of detection is imperfect, and it seems to be different for each type of wafer inspected. Although some logical associations can be made (highly reflective defects should be detected easily at low brightness, for exampl(*e*), parameter settings interact in unexpected ways, and their interactions depend on the type of wafer inspected and the particular defects of interest. For this reason, each new parameter setup is assumed to proceed from a state of nearly perfect ignorance.

A wafer type of special interest has been processed through metal etch, so it has a complex topography and a number of interesting real and nuisance defect types. The most important among them are:

1. Real Defects
   a. Metal corrosion (CORR): actual chemical corrosion observable on metal lines. This is a small but very high contrast defect, since the metal oxide appears black to the FIM camera.
   b. Black particle (BP): almost perfectly circular black particle
   c. Dark circle (DCIR(*c*): a dark circular donut.
   d. Blob (BLO(*b*): amorphous shape with ill-defined edges and highly variable size.
   e. Scratch (SCRATCH): very obvious catastrophic linear disruption of the wafer surface.

(*continues on facing page*)

2. Nuisance or False Defects
   a. Frost (FROST): harmless crystallization of the interlayer dielectric, resulting in a large but low-contrast nuisance.
   b. Interlayer dielectric bubble (BUBBLE): small but highly reflective nuisance.
   c. Metal grain (GRAIN): normal process variation during metal deposition sometimes produces metal crystals large enough to be detectable. These appear as a low-contrast network of lines within metal areas.

■ *Example 6.8: Optimizing a Visual Inspection System*

An optimum setting combination is needed for the following six predictors of inspection quality on the FIM:

1. Camera contrast.
2. Brightness of the inspection camera.
3. P12 adjusts sensitivity for straight edges.
4. P34 adjusts for corners.
5. P56 adjusts for voids.
6. P78 adjusts for protrusions.

The Initial Experiment
☐ Choose the predictors.
☐ Choose a centerpoint which is likely to produce useful information.
☐ Choose a step size for each predictor.
☐ Consider all important responses.
☐ Design the initial experiment.
☐ Run the initial experiment.

Assess the Experiment
☐ Evaluate the response at each of the experimental settings.

Plan the Next Experiment
☐ Determine if another experiment is necessary.
☐ Check for oscillation.
☐ "Reflect away" from the run with worst response of the last n+1 runs.
☐ Replace "worn-out" design points.

Verify the Chosen Optimum
☐ Assess process performance in the area around the optimum by running an experiment centered at that point.
☐ Run a production test.

Fig. 6.40 This checklist for simplex optimization is used to plan a sequence of experiments that require only one additional run after the initial experiment, and that systematically improve the desired response.

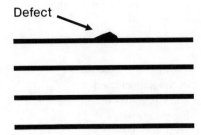

Fig. 6.41   P12 is a user-adjustable measurement parameter that modulates sensitivity to defects on straight edges, like the one shown here.

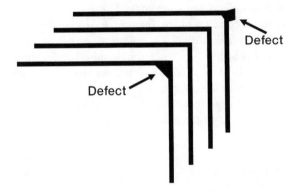

Fig. 6.42   P34 adjusts sensitivity to defects on the outside or inside of corners.

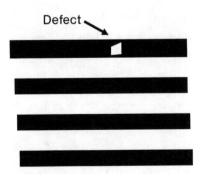

Fig. 6.43   P56 adjusts sensitivity to voids.

Since some defects are more important than others, a FOM is used to assess inspection quality:

Figure of merit =
5 times percent of dark circles detected +
3 times percent of corrosion detected +
2 times percent of black particles detected +
2 times percent of blobs detected +
1 times percent of scratches detected –
the percent of frost detected –
the percent of bubbles detected –
the percent of grain detected

Note that nuisance defects detract from the FOM.

So that inspection accuracy could be assessed, four typical wafers were selected from the production line and exhaustively examined. An enhanced inspection setup was used which cannot be used in production because it takes about 2.5 hours per wafer. The results of this enhanced inspection are used as the standard against which all other inspections are compared.

The initial simplex design shown below has seven runs, because there are six predictors. The centerpoint of the design was the manufacturer's recommended settings:

Contrast = 138
Brightness = 68
P12 = 85
P34 = 85
P56 = 55
P78 = 45

Step size for the design was chosen to be 11.

Run	Contrast	Brightness	P12	P34	P56	P78
1	133	65	83	83	54	44
2	145	65	83	83	54	44
3	138	75	83	83	54	44
4	138	68	92	83	54	44
5	138	68	85	92	54	44
6	138	68	85	85	62	44
7	138	68	85	85	55	52

After these first seven inspections were performed, the FOM was computed on the basis of their results. The computation is shown for the first setting combination here:

Defect	Seen	Actual	Weight	Adder
Frost	12	131	−1	−9.16
Bubble	2	185	−1	−1.08
Grain	0	15	−1	0.00
Corro	2	15	3	40.00
B. part.	3	18	2	33.33
D. circ.	1	6	5	83.33
Blob	29	29	2	200.00
Scratch	3	3	1	100.00
		FOM		446.42

The seven runs were sorted in ascending order of FOM (so the first point is the worst) and the settings for the next run were found:

Run	Contrast	Brightness	P12	P34	P56	P78	FOM
1	133	65	83	83	54	44	446
2	145	65	83	83	54	44	466
7	138	68	85	85	55	52	477
4	138	68	92	83	54	44	478
5	138	68	85	92	54	44	478
6	138	68	85	85	62	44	478
3	138	75	83	83	54	44	500
8	145.33	72.33	88.00	87.33	57.00	46.67	

Point 1 had the worst figure of merit, so it will be discarded; the other six points will be reused in following steps. Recall that settings for the new point (S_{NEW} in the equation below) are obtained by reflecting away from the worst point:

$$S_8 = \frac{S_2 + S_3 + S_4 + S_5 + S_6 + S_7}{3} - S_{worst}$$

The second iteration proceeds exactly as the first did, and produces settings for run number 9:

Run	Contrast	Brightness	P12	P34	P56	P78	FOM
2	145	65	83	83	54	44	466
7	138	68	85	85	55	52	477
4	138	68	92	83	54	44	478
5	138	68	85	92	54	44	478
6	138	68	85	85	62	44	478
3	138	75	83	83	54	44	500
8	145	72	88	87	57	47	500
9	133.33	74.67	89.67	88.67	58.00	47.67	

Continuing in this manner for 30 iterations resulted in an improvement in FOM from its initial value of 466 to about 620 at later setting combinations. The entire process is summarized graphically in Figure 6.44, with design points in order on the x-axis, and each setting plotted on the y-axis with the FOM. FOM values were divided by 5 so they could be plotted on the graph with the setting values.

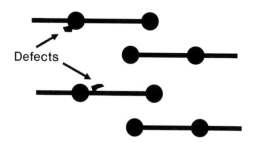

Fig. 6.44   P78 adjusts sensitivity for protrusions.

6.5   SUMMARY

Optimization experiments are among the last to be applied in the course of characterizing a process. They use more complex (and expensive) designs than any other type of experimentation, so it is important that they not be used until the process is well enough understood to benefit from their use.

The DSA method uses linear regression and common calculus to efficiently optimize using a few carefully chosen experimental runs. The method can exploit prior knowledge about observed or theoretical influences of predictors on responses to select suitable models and designs. Contour plots or other visual aids are essential to the intelligent application of the method.

The simplex method requires no knowledge of process relationships, and it works as well with six predictors as with two. It will almost always take longer to

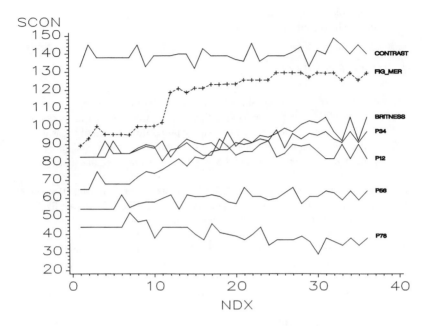

Fig. 6.45 The entire simplex optimization process for the FIM (Fast Inspection Machine) is represented on this plot. The run numbers are shown on the *x*-axis in the order the runs were made. The individual predictors are plotted on the *y*-axis along with the (scaled) figure of merit. Optimization seems to have reached a peak around run 25, and remained at that level for the rest of the experiment.

reach an optimum than the DSA, but in iteration-tolerant circumstances it is a very economical alternative.

The two methods presented in this chapter should be sufficient for most needs, but there are many other optimization topics which are worthy of mention.

- There are many optimization designs other than factorials and central composites with which to may be applied with the DSA: pentagonal, Box-Behnkin, D-optimal, and many other designs have been used. These designs, and the criteria used to choose among them, occupy a prominent place in the literature of experimental design. See Box and Draper (1987), or Myers et al. (1989) for further information.
- Evolutionary operation (EVOP) is a system of ongoing experimentation and optimization best applied to continuous processes; see Hunter and Kittrell (1966) or Box and Draper (1969) for more information.
- The presentation of the simplex method in this chapter represents only the elementary case; many variants and enhancements to the method have been developed. These variations involve, among others, changing step sizes depending on results, multiple reflection algorithms when more than a few predictors are involved, and elongation of some dimensions of the simplex. Anderson and McLean (1974) give an introduc-

tion to simplex designs; see Olsson (1974), or Olsson and Nelson (1975) for information on elaborated simplex designs.

CHAPTER 6 PROBLEMS

1. Which kind of experiment should be used in each of the following situations?
    (a) A gasoline manufacturer is investigating the effects of four factors on engine performance (mileage and cleanliness): proportion of ethanol in gasoline, proportion of unbranched alkanes, proportion of benzene, and proportion of branched alkanes. These are the only four constituents of this particular brand of gasoline.
    (b) A pizza lover has determined that two factors have a significant effect on pizza quality: cooking temperature, and cooking time. These factors are likely to interact, and their effect on the response might not be linear.
    (c) A tomato grower is investigating the effects of two different fertilizers on tomato growth rate and acidity. The fertilizers can be used alone, or in combination, and they are likely to have significant interaction effects.
    (d) A plasma etch engineer has chosen six factors that he thinks may affect etch rates. Nothing is known yet about their actual effects or interactions. Etching and measuring a single wafer takes about 18 minutes.
    (e) A metrology engineer in a semiconductor fabrication plant has just received a new machine that measures interlayer alignment. There are nine measurement parameters that the equipment manufacturer says will affect measurement speed—a critical outcome of the measurement process.

2. A response, Z, is known to be affected by two factors, X and Y, according to the following relationship:

$$Z = XY - 2X - 5Y + 10$$

    (a) Starting from ($X = 4$, $Y = 3$), what is the direction of steepest ascent?
    (b) If a step size of 1.5 were to be taken from this point, what would the centerpoint of an experiment centered one step away in the direction of steepest ascent?
    (c) Suppose that the response needed to be decreased, rather than increased. What is the direction of steepest descent from the starting point?
    (d) If a step size of 1.5 were to be taken from this point, what would the centerpoint be of an experiment centered one step away in the direction of steepest descent?

3. A response, Z, is known to be affected by two factors, X and Y, according to the following relationship:

$$Z = X^2 + XY - 2X - 5Y + 10$$

    (a) Starting from ($X = 4$, $Y = 3$), what is the direction of steepest ascent?
    (b) If a step size of 1.5 were to be taken from this point, what would the centerpoint of an experiment centered one step away in the direction of steepest ascent?

(c) Suppose that the response needed to be decreased, rather than increased. What is the direction of steepest descent from the starting point?

(d) If a step size of 1.5 were to be taken from this point, what would the centerpoint be of an experiment centered one step away in the direction of steepest descent?

4. A 2^2 factorial experiment with interaction was used to estimate a response surface; data from this experiment is shown below:

Factors		Responses	
X	Y		
−1	−1	18	20
−1	1	6	8
1	−1	12	14
−1	−1	4	6

(a) Estimate the parameters of this model.

(b) Determine the direction of steepest ascent from the centerpoint of this experiment.

(c) Suppose the model in problem 3 above was actually the right model. What would have been the DSA from (0, 0) using the more real model?

(d) How would the DSA differ for these two models if it were determined from the point (4, 3)?

(e) Does there appear to be lack of fit with the model determined in (b)?

5. A response, Z, is known to be affected by four factors, V, W, X and Y, according to the following relationship:

$$Z = X^2 + YV - 2X - 5Y + WX + 10$$

(a) Starting from ($V = 1$, $W = 2$, $X = 4$, $Y = 3$), what is the direction of steepest ascent?

(b) If a step size of 1.5 were to be taken from this point, what would the centerpoint be of an experiment centered one step away in the direction of steepest ascent?

(c) Suppose that the response needed to be decreased, rather than increased. What is the direction of steepest descent from the starting point?

(d) If a step size of 1.5 were to be taken from the original point, what would be the centerpoint of an experiment centered one step away in the direction of steepest descent?

(e) Determine the direction of steepest descent from the point determined in ((d), and find the point 1.5 units away in this direction.

(f) Make six projection contour plots (V by W with other factors at their original settings, V by X, etc.) to better understand the effects of these factors on Z.

6. A response, Z, is known to be affected by four factors, V, W, X, and Y, according to the following relationship:

$$Z = 3X^2 - W^2 + YV - 2X - 5Y + WX + 10$$

(a) Starting from ($V = 1$, $W = 2$, $X = 4$, $Y = 3$), what is the DSA?

(b) If a step size of 1.5 were to be taken from this point, what would be the centerpoint of an experiment centered one step away in the DSA?

(c) Suppose that the response needed to be decreased, rather than increased. What is the direction of steepest descent from the starting point?

(d) If a step size of 1.5 were to be taken from the original point, what would be the centerpoint of an experiment centered one step away in the direction of steepest descent?

(e) Determine the direction of steepest descent from the point determined in (d), and find the point 1.5 units away in this direction.

(f) Make six projection contour plots (V by W with other factors at their original settings, V by X, etc.) to better understand the effects of these factors on Z.

7. A tea lover has asked that you help him design an experiment to optimize tea strength to a desired target. Two factors are involved in this experiment: brewing temperature and steeping time.

(a) Design a simple factorial experiment for this purpose; center the settings at (80°C, 4 minutes), and select a step size for each factor based on your understanding of the physical situation.

(b) Explain your choices to be the tea lover.

(c) Design a central composite experiment for this purpose using the same instructions as in part (a).

(d) Explain the benefits of this more complex experiment to the tea lover.

8. A pizza lover has asked that you help him design an experiment to optimize pizza crust crispiness to a desired target. Two factors are involved in this experiment: cooking temperature and cooking time.

(a) Design a simple factorial experiment for this purpose; center the settings at (375°F, 18 minutes), and select a step size for each factor based on your understanding of the physical situation.

(b) Explain your choices to the pizza lover.

(c) Design a central composite experiment for this purpose using the same instructions as in part (a).

(d) Explain the exact sequence of runs and treatment combinations in this experiment in a written memorandum to the pizza lover.

9. I have decided to use a central composite experiment to determine the influence of two factors on the time it takes me to get to work: front bicycle tire pressure, and back tire pressure. I live 4 miles from work, and the ride usually takes about 18 minutes with the tire pressure I currently use (35 lb for both tires).

(*a*) Design a central composite experiment for this purpose; center the settings at the current tire pressures, and select a step size for each factor based on your understanding of the physical situation.

(*b*) I don't want to have to look up the tire pressure I need every morning, or change pressures very often, so randomization will be a nuisance. Is randomization really necessary here?

(*c*) I want to get this experiment done as quickly as possible; is it really necessary to replicate this experiment?

10. Suppose that I also wanted to include a "breakfast" factor in the experiment above; this factor takes the −1 value if I skip breakfast, and a +1 value if I eat breakfast.
    (a) What kind of experiment would be most appropriate for this problem?
    (*b*) Describe an experiment that would suit my needs, but take no more than 8 days.

11. The low temperature oxide (LTO) thickness target for a new process is 1150 Å. A factorial experiment centered at the current reaction temperature (620) and reaction time (425) was conducted to optimize the process. Data from the two replicates of this experiment is shown below:

TEMP	TIME (min)	THICK (Å)
600	400	1138
600	400	1156
600	450	1181
600	450	1192
640	400	1155
640	400	1126
640	450	1161
640	450	1166

(*a*) Use analysis of variance to test the significance of both predictors and their interaction.

(*b*) Estimate the model parameters, and determine the direction of steepest ascent or descent—whichever is most appropriate.

(*c*) Assume that the step size will be the same as it was in the previous experiment (25 for reaction time and 20 for reaction temperatur(*e*) and choose a center-point for the next experiment.

(*d*) Use the pure error lack of fit test to test the appropriateness of the model.

12. Another LTO thickness optimization experiment was centered at (620, 425), and a central composite design was used. Results from the experiment are shown below:

TEMP	TIME	THICK
600	400	1138
600	450	1193
640	400	1144
640	450	1192
592	425	1177
648	425	1148
620	390	1113
620	464	1172
620	425	1166
620	425	1146
620	425	1164
620	425	1168
620	425	1162
600	400	1140
600	450	1196
640	400	1125
640	450	1177
592	425	1165
648	425	1168
620	390	1147
620	464	1186
620	425	1182
620	425	1163
620	425	1151
620	425	1154
620	425	1174

(*a*) Use analysis of variance to test the significance of all the effects in the model.

(*b*) Estimate the model parameters, and determine the direction of steepest ascent or descent, whichever is most appropriate.

(*c*) Assume that the step size will be the same as it was in the previous experiment and choose a centerpoint for the next experiment.

(*d*) Use the pure error lack of fit test to test the appropriateness of the model.

13.   The resist thickness target for a new process is 11,250 Å. A factorial experiment centered at the current coat temperature (32) and spin time (95) was conducted to optimize the process. Data from the two replicates of this experiment is shown below:

TEMP	TIME	THICK
30	90	11,228
30	90	11,246
30	100	12,203
30	100	12,214
34	90	11,221
34	90	11,192
34	100	12,159
34	100	12,165

(a) Use ANOVA to test the significance of both predictors and their interaction.
(b) Estimate the model parameters, and determine the direction of steepest ascent or descent, whichever is most appropriate.
(c) Assume that the step size will be the same as it was in the previous experiment (95 for time and 32 for temperatur(e) and choose a centerpoint for the next experiment.
(d) Use graphical methods to assess model fit.
(e) Use the pure error lack of fit test to test the appropriateness of the model.

14. Another resist thickness optimization experiment was centered at (32,95), and a central composite design was used. Results from the experiment are shown below:

TEMP	TIME	THICK
30	90	11,228
30	100	12,216
34	90	11,209
34	100	12,190
29.2	95	11,654
34.8	95	11,592
32	88	11,084
32	102	12,447
32	95	11,627
32	95	11,607
32	95	11,625
32	95	11,629
32	95	11,623

(*a*) Use ANOVA to test the significance of all the effects in the model.

(*b*) Estimate the model parameters, and determine the direction of steepest ascent or descent, whichever is most appropriate.

(*c*) Assume that the step size will be the same as it was in the previous experiment and choose a centerpoint for the next experiment.

(*d*) Use the pure error lack of fit test to test the appropriateness of the model.

15. The uniformity of resist thickness is also an important process outcome. The standard deviation of resist thickness was measured in the central composite experiment above, with results as shown:

TEMP	TIME	THICK STD
30	90	66
30	100	64
34	90	47
34	100	38
29.2	95	77
34.8	95	15
32	88	43
32	102	21
32	95	50
32	95	30
32	95	48
32	95	52
32	95	46

(*a*) Use ANOVA to test the significance of all the effects in the model on the square root of resist thickness standard deviation.

(*b*) Estimate the model parameters, and determine the direction of steepest ascent or descent, whichever is most appropriate. (Uniform resist thickness is good; nonuniform resist thickness is bad.)

(*c*) Assume that the step size will be the same as it was in the previous experiment and choose a centerpoint for the next experiment.

(*d*) Use the pure error lack of fit test to test the appropriateness of the model.

16. A five-predictor central composite design based on the 2^{5-1} fractional factorial was used to optimize complimentary metal oxide semiconductor (CMOS) transistor performance. The factors involved were: polisilicon CD (PCD), field oxide CD (FCD), p-well implant dose (PDOSE), p-well implant energy (PKEV), and n-tip implant dose (NTIP). The responses were the maximum frequency at which the device passed all functional tests (SP), and the current consumed in stand-by mode (IDOWN). High-operation frequency and low stand-by current are both desirable. Standby current greater than 70 mA is intolerable. Data from the experiment is shown below:

PCD	FCD	PDOSE	PKEV	NTIP	SP	IDOWN
2.50	4.1	11.70	23	4.22	71	16.6
2.50	4.1	11.70	25	4.18	71	16.4
2.50	4.1	11.80	23	4.18	72	4.0
2.50	4.1	11.70	23	4.22	67	12.1
2.50	4.9	11.70	23	4.18	71	55.3
2.50	4.9	11.70	25	4.22	72	59.0
2.50	4.9	11.80	23	4.22	69	49.9
2.50	4.9	11.80	25	4.18	73	47.4
3.00	4.1	11.70	23	4.18	70	14.5
3.00	4.1	11.70	25	4.22	73	16.9
3.00	4.1	11.80	23	4.22	70	9.3
3.00	4.1	11.70	23	4.18	71	12.2
3.00	4.9	11.70	23	4.22	69	59.9
3.00	4.9	11.70	25	4.18	70	59.1
3.00	4.9	11.80	23	4.18	74	48.8
3.00	4.9	11.80	25	4.22	69	48.6
2.25	4.5	11.75	24	4.20	45	35.7
3.25	4.5	11.75	24	4.20	45	29.2
2.75	3.7	11.75	24	4.20	88	4.0
2.75	5.3	11.75	24	4.20	89	77.1
2.75	4.5	11.65	24	4.20	91	38.3
2.75	4.5	11.85	24	4.20	92	22.2
2.75	4.5	11.75	22	4.20	91	33.2
2.75	4.5	11.75	26	4.20	93	33.1
2.75	4.5	11.75	24	4.16	92	32.1
2.75	4.5	11.75	24	4.24	93	30.7
2.75	4.5	11.75	24	4.20	95	29.9
2.75	4.5	11.75	24	4.20	87	28.7
2.75	4.5	11.75	24	4.20	93	32.4
2.75	4.5	11.75	24	4.20	85	32.2
2.75	4.5	11.75	24	4.20	87	32.4
2.75	4.5	11.75	24	4.20	93	34.8

(*a*) Use ANOVA to test the significance of all the effects in the model on the responses. Make interaction plots for any significant interactions.

(*b*) Estimate the model parameters, and determine the direction of steepest ascent or descent—whichever is most appropriate—for each response.

(*c*) Assume that the step size will be the same as it was in the first experiment and choose a centerpoint for the next experiment. Take both responses into consideration.

(*d*) Use projection contour plots to understand the surface predicted by the model.

(*e*) Use the pure error lack of fit test to test the appropriateness of the model.

17. Brewing and tasting a cup of tea is a very quick process, so a simplex experiment will be used to optimize tea strength. Design a simplex experiment with two fac-

tors: brewing temperature and steeping time. The experiment should be centered at (85°C, 6 minutes). Step size for temperature is 2°; step size for time is 15 seconds.

18. The results from a tea brewing simplex experiment with two factors (temperature and steeping tim(e) are shown below. Assume that stronger tea is desirable, and determine the temperature and time for the next run of this experiment.

TEMP	TIME	STR
78.5	223	126.6
81.5	223	128.7
80	283	138.1

★ 19. Photoresist is so hardened by certain ion implants that it must be *ashed* before it can be stripped with sulfuric acid. Ashing is an aggressive plasma etch specifically formulated to disintegrate organic polymers. Important outcomes of the process are the polymer etch rate (bigger is better), and the substrate etch rate (smaller is good). The substrate etch rate cannot be allowed to exceed 4 Å/s. The starting point and step size for the six process parameters involved are shown below:

Factor	Centerpont	Step Size
Plasma energy	2800 W	100 W
Temperature	240°C	50°C
SO_2 flow	20 SCCM	1 SCCM
Cl_2 flow	80 SCCM	2 SCCM
Pressure	26 torr	1 torr
Microwave wavelength	10 mm	0.01 mm

(*a*) Determine the settings for the initial simplex experiment, which will require seven runs.

(*b*) Suppose that third setting produced the least desirable results; what would be the next setting combination in the sequence?

(*c*) How are you going to balance the need to increase polymer etch rate with the constraint on substrate etch rate?

★ 20. Suppose, that in the experiment above, polymer etch rate were related to the six process factors as follows:

$$\text{Rate }(E,T,S,C,W) = 80\,f_1(E)f_2(T)f_3(c/s)f_4(W)$$

where:

$$f_1(E) = 0.96 \quad \text{for} \quad E \le 2200$$
$$= 0.96 + 0.00003636\,(E - 2200) \quad \text{for} \quad 2200 < E \le 3300$$
$$= 1 \quad \text{for} \quad E > 3300$$

and

$$f_2 = 0.98 \quad \text{for} \quad T \le 200$$
$$= 0.98 + 0.0014T \quad \text{for} \quad 200 < T \le 250$$
$$= 1.05 \quad \text{for} \quad T > 250$$

and

$$f_3\left(\frac{C}{S}\right) = e^{-5\left(\frac{C}{S}-3.2\right)^2} + 0.40e^{-5\left(\frac{C}{S}-4.8\right)^2}$$

and

$$f_4(P) = 1 \quad \text{for} \quad P \le 19$$
$$= 1.00 - 0.04(P - 19) \quad \text{for} \quad 19 < P \le 26$$
$$= 0.72 \quad \text{for} \quad P > 26$$

In these equations, E = energy, T = temperature, C/S is the ratio of chlorine flow to sulfur dioxide flow, and P is pressure.
(a) Evaluate the polymer etch rate function at each of the seven treatment combinations of the experiment determined in (a) of the previous problem.
(b) Continue the sequence of simplex experimentation steps by reflecting away from the worst treatment combination, and then evaluating the response at the new point.
(c) Have you found an optimum? Do you believe this is a local or a global optimum?

TECHNICAL NOTES

TECHNICAL NOTES FOR CHAPTER 3

Table 3.1 was produced by iteratively increasing sample size (n) until each given set of experimental conditions was met: number of treatments (p), number of blocks (q), and difference to be found (Δ). Δ is derived from δ: the difference at which beta risk is to be controlled in the original scale of the data. For the model without interaction, Δ is scaled by σ—the lowest component of experimental error, uncontaminated by block variance: $\Delta = \delta/\sigma$.

For the model without interaction, the F-test has $p - 1$ degrees of freedom for the numerator (DFN), and $pqn - p - q + 1$ degrees of freedom for the denominator (DFD). Under H_0, the test statistic has a (central) F-distribution, so the critical value is $F_{1 - \alpha, \text{DFN, DFD}}$. Recall that, by definition,

$$P\{F_{DFT, DFE} \leq F_{1 - \alpha, DFT, DFE}\} = 1 - \alpha$$

where $F_{\text{DFN, DFD}}$ is an F-distributed random variable with DFN and DFD degrees of freedom.

Under H_1, the test statistic has a noncentral F-distribution with noncentrality parameter $\lambda = nq\Delta^2/2$. To ensure that beta risk is controlled, sample size was increased until:

$$P\{F_{DFT, DFEl, \lambda} \leq F_{1 - \alpha, DFT, DFE}\} \leq \beta$$

289

where $F_{DFN, DFD, \lambda}$ is a noncentral F-distributed random variable with DFN and DFD degrees of freedom, and noncentrality parameter λ.

The model with interaction requires the same computations, but with DFD = $(p-1) * (q-1)$, and a different interpretation for Δ. In this case, the expected mean square (EMS) for the treatment factor is:

$$E(MST) = \sigma^2 + nq \, \frac{\sum\limits_{i=1}^{p} \tau_i^2}{p-1} + n\sigma_{\tau\beta}^2$$

so the correct denominator is not σ, but the expression:

$$\sqrt{\sigma^2 + n\sigma_{\tau\beta}^2}$$

ANSWERS TO PROBLEMS

CHAPTER 2 ANSWERS TO PROBLEMS

1(a) 103

1(b) Some form of randomization is absolutely necessary here because of the influence player fatigue and learning on the response over time. Small groups of kicks could be grouped by shoe to complete the experiment in a single day.

5(a) ANOVA may be valid even if these are counts—if the residuals are normally distributed, the model is valid.

5(b) The F statistic from ANOVA is only 0.94, with a corresponding p-value of 0.4425. There does not appear to be a significant difference between the SEMs.

5(c) The sample chi-squared statistic is 6.1323, but the 95th percentile of the chi-squared distribution with three degrees of freedom is 7.81. There is no significant difference between the instruments.

7(a) The machines are significantly different.

7(b) Machine 3 has a significantly higher mean than the other two machines, as indicated by the Tukey grouping below:

Tukey Grouping	Mean	N	MACH
A	0.63860	5	3
B	0.60120	5	1
B			
B	0.59920	5	2

11(a) The means are not significantly different.

11(b) An estimate of the within-group standard deviation can be obtained from the ANOVA table: 5.4375. Detecting a 2% difference in yield would be prohibitively expensive; Detecting a 5.44% difference would require 27 lots to be run on each machine, for a total of 81 lots.

CHAPTER 3 ANSWERS TO PROBLEMS

2(a) TA appears to have an insignificant effect on the scores: the F statistic is only 0.15.

2(b) TA now appears to be significant: the F statistic is 13.37, which has a p value of 0.0001.

4(b) Brand does not have a significant effect on head height, even if the interactive model is used.

4(d) An estimate of within-group standard deviation (for the head height response) is obtained from the ANOVA table: 1.2572. The desired difference is about 0.80 standard deviations. Using Table 9.1 reveals that, if only five bottles from each batch are to be sampled, then ten batches must be brewed for this experiment. If 10 bottles from each batch could be sampled, then only five batches must be brewed. (These sample sizes are both for the noninteractive model.)

5(b) Location 1 provides no useful information; leaving this data in the analysis artificially enhances the significance of the factors.

5(c) The interactive model indicates that the PAINT effect is not significant, although the model without interaction indicates that PAINT is significant. The model with interaction is the correct model, because the PAINT*LOC interaction is significant.

9(a) The testers are equivalent.

9(b) Yes.

CHAPTER 4 ANSWERS TO PROBLEMS

1(a) This is a 2^2 factorial, replicated twice.

1(b) The point estimate for the interaction effect is –6.5.

1(d)　Temperature is the only significant effect.

1(e)　ANOVA provides an estimate of within-group standard deviation: 5.44. The desired difference is 1.8373 standard deviations, and the probability of detecting this difference with an experiment this size is 0.29324.

1(f)　Use Table 10.1 with $\Delta = 1.0$. The experiment would have to be replicated 12 times.

3(b)　The TEMP*TOP, TEMP*TIME, and the TEMP*TIME*TOP effects seem significant in plots.

3(c)　The only significant effects are TEMP*TOP and TEMP*TIME. The p value for the three-way interaction is 0.0922.

3(d)　Pepperoni was used for all of the centerpoints. If the crispiness variance is different for pepperoni than for artichoke hearts, then the validity of the statistical tests may be in question.

7(a)　Effects estimates are given below—"R" is used for the BREATH factor. The effect of HEAD ("H") is clearly significant.

Effect	Estimate
RFBH	−7.25
RB	−5.25
FB	−3.75
RH	−3.75
RFH	−2.75
RFB	−1.25
F	−0.25
R	0.75
RF	1.25
FBH	1.25
RBH	2.25
FH	3.75
B	5.25
BH	7.75
H	18.75

7(b)　Be sure to keep your eye on the ball when you serve.

7(d)　Use a 2^{6-2} fractional factorial experiment—see Chapter 5.

9(a)　AGE, and the COSTUME*AGE interaction are both significant.

9(b)　The age of the children in the experiment is confounded with their sex, and there is also likely to be interaction between costume type and sex.

CHAPTER 5 ANSWERS TO PROBLEMS

1(a)　Yes.

1(b)　No: the A, B, and C effects are already known to be significant so a screening experiment would be superfluous, and it would fail to detect interactions which seem likely in this context. The remaining factors (D through K) are good subjects for a screening experiment.

1(c) This response is categorical, so screening experiments as explained here would not apply. If the crunch could be suitably quantified, then the ANOVA assumptions would be valid.

1(d) The top 11 suspects would be good subjects for the 2^{11-7} screening experiment. A more expedient method might be to use a 23-factor Plackett-Burman experiment to test all the factors at once.

1(e) Capacitor width and capacitor length determine the area of the capacitor, which is directly proportional to the capacitance—these factors are known significant influences and they interact, so screening experiments are inappropriate.

10(a) Exposure and focus are significant factors. Due to the resolution of this design, interactions cannot be estimated.

15(a) Effects estimates are shown below with Y for the hydration effect and R for the breath effect. The HEAD and HEAD-BACK interaction both appear to be significant.

Effect	Estimate
YB	–4.375
YH	–3.375
FB	–3.125
Y	–2.875
R	–2.125
RB	–2.125
RH	–1.125
F	–0.125
RY	0.875
YF	0.875
RF	2.125
FH	2.875
B	5.625
BH	7.125
H	18.625

15(c) Significant effects now are: HEAD (*p* value 0.0012), and HEAD*BACK (*p* value 0.0539 is nearly significant).

15(d) Use a 2^{6-2} design with block as the sixth factor, and choose factors in such a way that the confounded interactions are very likely to be unimportant.

16(a) Reverse the sign of every treatment in the original design, so the first run would be:

$$(A = 1, B = 1, C = 1, D = -1, E = -1, F = -1)$$

16(b) Rerun the original experiment changing only the signs of factor C, so the first run would be:

$$(A = -1, B = -1, C = 1, D = 1, E = 1, F = 1)$$

20(a) Significant factors are: A, B, D, F, and G.

CHAPTER 6 ANSWERS TO PROBLEMS

1(a) This is a mixture experiment because the factor settings must add to a constant—these require special designs that were not shown in this book.

1(b) Use a 2^2 factorial or a central composite design.

1(d) Use a screening design to see which of the factors have some effect, then consider a simplex design to optimize using the few significant factors.

1(e) Use a simplex design involving all nine factors.

2(a) The DSA from this point is along the $(X = -1, Y = -1)$ vector.

2(b) $(X = -1.06, Y = -1.06)$

7(c) Use the same factorial points, and add five centerpoints and axial points at a distance of 1.414 steps from the center. If a 5-degree step size had been selected for temperature, then the two temperature axial points would be: (TEMP = 87, TIME = 4) and (TEMP = 73, TIME = 4).

7(d) This experiment will be able to detect curvature which was invisible to the simple factorial.

17. The three runs required for the initial experiment are:

TEMP	TIME
84	366 sec
86	366 sec
84	371 sec

18. Temperature = 83.5, Time = 283.

REFERENCES

Anderson, V. L., and McLean, R. A. *Design of Experiments,* Marcel Dekker, New York, 1974.

Booth, K. H. V., and Cox, D. R. "Some Systematic Supersaturated Designs," *Technometrics,* Vol. 4, 489–495 (1962).

Box, G. E. P., and Draper, N. R. *Evolutionary Operation,* Wiley, New York, 1969.

Box, G. E. P., and Draper, N. R. *Empirical Model-Building and Response Surfaces,* Wiley, New York, 1987.

Box, G. E. P., Hunter, W. G., and Hunter, J. S. *Statistics for Experimenters,* Wiley, New York, 1978.

Coleman, D. E., and Montgomery, D. C. "A Systematic Approach to Planning for a Designed Industrial Experiment," *Technometrics,* Vol. 35, No. 1, 1–27 (1993).

Cornell, J. A. *Experiments with Mixtures: Designs, Models, and the Analysis of Mixture Data,* Wiley, New York, 1981.

Drain, David C., Gough, Andrew M. "Applications of the Upside-Down Normal Loss Function", *IEEE Transactions on Semiconductor Manufacturing,* Volume 9 Number 1, pp. 143–145 (1996).

Hollander, M., and Wolfe, D. A. *Nonparametic Statistical Methods,* Wiley, New York, 1973.

Hunter, W. G., and Kittrell, J. R. "Evolutionary Operation: A Review," *Technometrics,* Vol. 8, 389–397 (1966).

Lenth, R. "Quick and Easy Analysis of Unreplicated Factorials," *Technometrics,* Vol. 31, 469–473 (1989).

Lorenzen, T. J., and Anderson, V. L. *Design of Experiments, A No-Name Approach,* Marcel Dekker, New York, 1993.

Miller, R. G., Jr. *Simultaneous Statistical Inference,* 2nd ed., Springer, New York, 1981.

Montgomery, D. C. *Design and Analysis of Experiments,* 2nd ed., Wiley, New York, 1984.

Myers, R. H., Khuri, A. I., and Carter, W. H., Jr. "Response Surface Methodology: 1966–1988," *Technometrics,* Vol. 31, 137–157 (1989).

Nelson, L. S. "Comparison of Poisson Means: The General Case," *Journal of Quality Technology,* Vol. 19, No. 4, 173–179 (1987).

Neter, J., Wasserman, W., Kutner, M. H. *Applied Linear Statistical Models,* 2nd ed., Irwin, Illinois 1985.

Olsson, D. M., "A Sequential Simplex Program for Solving Minimization Problems", *Journal of Quality Technology,* Vol. 6, 53–57 (1974).

Olsson, D. M., and Nelson, L. S. "The Nelder-Mead Simplex Procedure for Function Minimization," *Technometrics,* Vol. 17, Number 1, 45–51 (1975).

Papazian, C. *The New Complete Joy of Home Brewing,* Avon, New York, 1991.

Pignatiello, J. J., Jr., and Ramberg, J. S. "Top Ten Triumphs and Tragedies of Genichi Taguchi," *Quality Engineering,* Vol. 4, No. 2, 211–226 (1991–1992).

Taguchi, G., *System of Experimental Design,* Vol. 1 and 2, Unipub/Kraus, White Plains, N Y, (1987).

Thomas, G. B., and Finney, R. L. *Calculus and Analytic Geometry,* Addison-Wesley, Reading, MA, 1988.

Watson, G. S. "A Study of the Group Screening Method," *Technometrics,* Vol. 3, No. 3, 371–388 (1961).

Westfall, P. H., and Young, S. S. *Resampling-Based Multiple Testing,* Wiley, New York, 1993.

APPENDIX A: THE STATISTICAL ANALYSIS SYSTEM (SAS)

The purpose of this supplement is to provide readers with a means to apply SAS to accomplish the tasks in the text. Some basic knowledge of the software is assumed; this can be obtained from the excellent manuals and classes offered by SAS institute.

The appendix consists of a series of programs, with input data sets and output listings. Programs are numbered to correspond to the chapters in which they would likely be used first: Program S2_1.SAS is the first program related to Chapter 2, for example. Programs and input data are available at the International Thomson Publishing web site, www.thomson.com. This appendix contains no listings.

Programs in this appendix were written for use on an Intel 80386 machine using MS-DOS, so file names reflect that environment. They can be adapted to run in other environments by changing input file names, and if necessary, changing graphics options to correspond to available peripherals.

PROGRAM S_1.SAS: ONE-WAY ANALYSIS OF VARIANCE

This program does analysis of variance on a single treatment factor, and produces two different displays of the Tukey means comparison procedure results.

```
OPTIONS LS=68 PS=48;
/*

        S2_1.SAS Standard ANOVA

    1.  Read raw data
    2.  Do analysis of variance
    3.  Do Tukey Means comparisons two ways

*/
DATA D1;
        INFILE 'S2_1.DAT' PAD;
        LENGTH TUBE $ 3;
        INPUT T R;
        IF T=1 THEN TUBE='13A';
        IF T=2 THEN TUBE='13B';
        IF T=3 THEN TUBE='13C';
        IF T=4 THEN TUBE='13D';

*       Analysis of variance;
PROC GLM DATA=D1;
        CLASS TUBE;
        MODEL R=TUBE;
        MEANS TUBE/TUKEY CLDIFF;
        MEANS TUBE/TUKEY LINES;
```

PROGRAM S2_2.SAS: ANALYSIS OF POISSON DATA

This program uses the binomial distribution to test the equivalence of two defect densities. The areas sampled can be different.

The second part of the program uses the chi-squared distribution to make a similar test regarding four populations.

```
OPTIONS LS=68 PS=48;
/*

        S2_2.SAS—analysis of Poisson data
    1.  Read and test two Poisson samples with different areas sampled
    2.  Read and test four Poisson samples with different areas sampled

*/
DATA D1;
        INFILE 'S2_2A.DAT';
        INPUT N1 A1 N2 A2;
```

```
        P0=A1/(A1+A2);
        P=N1/(N1+N2);
        PVAL=PROBBNML (P0, N1+N2, N1); OUTPUT;
        PVAL=1−PROBBNML (P0, N1+N2, N1−1); OUTPUT;
PROC PRINT DATA=D1;
        TITLE 'Binomial test on two samples';
DATA D2;
        INFILE 'S2_2B.DAT';
        INPUT N A;
PROC MEANS DATA=D2;
        VAR N A;
        OUTPUT OUT=D3 SUM=SUM_N SUM_A;
        TITLE 'Chi-squared test on Poisson Data';
        DATA D4;
        SET D3 D2;
        RETAIN CHISQ 0 N_TOT 0 AREA_TOT 0;
        IF _N_=1 THEN DO;
             N_TOT=SUM_N;AREA_TOT=SUM_A;
        END;
        ELSE DO;
             E=N_TOT*A/AREA_TOT;
             CONTR=(N−E)**2/E;
             CHISQ=CHISQ+CONTR;
             OUTPUT;
        END;
        IF _N_=5 THEN DO;
             P=CINV(0.95, 3);
             OUTPUT;
        END;
PROC PRINT DATA=D4;
        VAR N A E CONTR CHISQ N_TOT AREA_TOT P;
```

PROGRAM S2_3.SAS: NONPARAMETRIC ANOVA

This program uses the Kruskal-Wallis test to do a nonparametric analysis of variance on data from nonnormal populations. The populations compared must be continuous and symmetric about their medians.

```
OPTIONS LS=68 PS=48;
/*
        S2_3.SAS—Nonparametric ANOVA
   1.   Read data from nonnormal population spread across several tubes
   2.   Do a nonparametric test for equality of medians
```

```
*/
DATA D1;
        INFILE 'S2_3.DAT' PAD;
        LENGTH TUBE $ 3;
        INPUT TUBE RHO;
PROC NPAR1WAY DATA=D1 WILCOXON;
        CLASS TUBE;
        VAR RHO;
        TITLE 'Kruskal-Wallis Test';
```

PROGRAM S3_1.SAS: ANOVA WITH BLOCKING

This program performs several different analyses of blocked experiments. When
the interaction term is included in a blocked model, the program must be prompted
to perform the correct hypothesis test for the treatment factor as follows:

$$TEST\ H = IMP\ E = LOT*IMP;$$

This forces an *F*-ratio to be computed with IMPlanter sum of squares in the numerator, and the lot by implanter interaction in the denominator.

```
OPTIONS LS=68 PS=48;
/*

        S3_1.SAS—analysis of blocked experiment
                with and without interaction term
    1.  Read raw data
    2.  do anova with interaction term
    3.  do anova without interaction term
    4.  Set up graphics device
    5.  beta risk curves for unblocked model
    6.  compute effects estimates
    7.  interaction plots
    8.  assumption checks—residuals histogram
    9.  assumption checks—timeplot of residuals

*/
*       Read raw data;
DATA D1;
        INFILE 'S3_1.DAT';
        INPUT LOT IMP VT WAFER RUNORD;
*       do anova with interaction term;
```

```
PROC GLM DATA=D1;
      CLASS LOT IMP;
      MODEL VT=LOT IMP LOT*IMP/SOLUTION;
      TEST H=IMP E=LOT*IMP;
      MEANS IMP/TUKEY E=LOT*IMP ETYPE=3;
      MEANS IMP/TUKEY;
      TITLE 'Interaction Model';
*     do anova without interaction term;
PROC GLM DATA=D1;
      CLASS LOT IMP;
      MODEL VT=LOT IMP/SOLUTION;
      MEANS IMP/TUKEY;
      TITLE 'No Interaction Model';
*     Set up graphics device;
GOPTIONS DEVICE='HP7475'
      VPOS=48 HPOS=60 ASPECT=1 HBY=1.5
      HTITLE=5 FTITLE=SIMPLEX
      HTEXT=2 FTEXT=SIMPLEX
      ROTATE=LANDSCAPE
      NODISPLAY
      GSFMODE=REPLACE
      GSFNAME=GFIL;

*     beta risk curves for unblocked model;
DATA D4;
      CV=FINV(0.95, 3, 18);
      DO DEL=0.0 TO 3.4 BY .025;
            DELTA=DEL*0.071062;
            BETA=PROBF(CV, 3, 18, 3*DEL**2);
            OUTPUT;
      END;
SYMBOL1 COLOR=BLACK I=SPLINE V=NONE;
FILENAME GFIL 'S3_1A.GSF';
PROC GPLOT DATA=D4;
      PLOT BETA*DELTA=1/VAXIS=0 TO 1 BY .1
            HAXIS=0 TO 0.25 BY 0.05; RUN;

*     compute effects estimates;
PROC SUMMARY DATA=D1;
      CLASS IMP;
      VAR VT;
      OUTPUT OUT=D2 MEAN=M_VT;
DATA D3;
      SET D2;
```

```
       RETAIN GRANMEAN;
       IF _TYPE_=0 THEN GRANMEAN=M_VT;
       ELSE DO;
              IMP_EFF=M_VT–GRANMEAN;
              OUTPUT;
       END;
PROC PRINT DATA=D3;
       TITLE 'Implanter Effects Estimates';
PROC SUMMARY DATA=D1;
       CLASS LOT;
       VAR VT;
       OUTPUT OUT=D2 MEAN=M_VT;
DATA D3;
       SET D2;
       RETAIN GRANMEAN;
       IF _TYPE_=0 THEN GRANMEAN=M_VT;
       ELSE DO;
              LOT_EFF=M_VT–GRANMEAN;
              OUTPUT;
       END;
PROC PRINT DATA=D3;
       TITLE 'Lot Effects Estimates';
*      interaction plots;
PROC SUMMARY DATA=D1 NWAY;
       CLASS LOT IMP;
       VAR VT;
       OUTPUT OUT=D2 MEAN=M_VT;
TITLE ";
SYMBOL1 COLOR=BLACK I=JOIN V=SQUARE;
SYMBOL2 COLOR=BLACK I=JOIN V=TRIANGLE;
SYMBOL3 COLOR=BLACK I=JOIN V=DIAMOND;
SYMBOL4 COLOR=BLACK I=JOIN V=CIRCLE;
FILENAME GFIL 'S3_1B.GSF';
DATA D7;
       SET D2;
       IF IMP=1 THEN DO;
              VT=M_VT;VT2=.;VT3=.;VT4=.;OUTPUT;END;
       IF IMP=2 THEN DO;
              VT=.;VT2=M_VT;VT3=.;VT4=.;OUTPUT;END;
       IF IMP=3 THEN DO;
              VT=.;VT2=.;VT3=M_VT;VT4=.;OUTPUT;END;
       IF IMP=4 THEN DO;
              VT=.;VT2=.;VT3=.;VT4=M_VT;OUTPUT;END;
```

```
PROC GPLOT DATA=D7;
      PLOT VT*LOT=1 VT2*LOT=2 VT3*LOT=3 VT4*LOT=4/
      OVERLAY HMINOR=0;
*     assumption checks—residuals histogram;
PROC GLM DATA=D1;
      CLASS LOT IMP;
      MODEL VT=LOT IMP;
      MEANS IMP/TUKEY;
      OUTPUT OUT=D5 R=RESID;
FILENAME GFIL 'S3_1C.GSF';
PROC GCHART DATA=D5;
      VBAR RESID/ SPACE=0 MIDPOINTS=–.15 TO .15 BY .05;RUN;


*     assumption checks—timeplot of residuals
SYMBOL1 COLOR=BLACK I=JOIN V=SQUARE;
FILENAME GFIL 'S3_1D.GSF';
PROC SORT DATA=D5;BY RUNORD;
PROC GPLOT DATA=D5;
      PLOT RESID*RUNORD;RUN;
```

PROGRAM S3_2.SAS: CROSSOVER DESIGN ANALYSIS

The crossover design is analyzed by adding a factor to the model to account for the order in which subjects (wafers in this case) are tested. This ORDER factor is used like any other blocking factor.

```
OPTIONS LS=68 PS=48;
/*

      S3_2.SAS—crossover design

   1. Read data from a crossover design—the order of the test is either one or
      two, and each wafer is tested twice.
   2. Analyze in ANOVA using ORDER as a blocking factor.

*/
DATA D1;
      INFILE 'S3_2.DAT';
      INPUT TESTER ORDER WAFER YIELD;
PROC GLM DATA=D1;
      CLASS TESTER ORDER WAFER;
      MODEL YIELD=TESTER ORDER WAFER;
```

PROGRAM S3_3.SAS: LATIN SQUARE ANALYSIS

Two blocking factors, WEEK and SUBJ, are included in this Latin square design. The treatment factor of interest is the brand of headgear (HEADGR). No estimates of interactions can be made.

```
OPTIONS LS=68 PS=48;
/*
        S3.3SAS—Latin square analysis
    1.  Read data with two blocking factors (WEEK and SUBJ) and one treatment
        factor (HEADGR)
    2.  Do ANOVA

*/
DATA D1;
        INFILE 'S3_3.DAT';
        INPUT WEEK SUBJ HEADGR $ SCORE;
PROC GLM DATA=D1;
        CLASS WEEK SUBJ HEADGR;
        MODEL SCORE=WEEK SUBJ HEADGR;
```

PROGRAM S4_1.SAS: ANALYSIS OF FACTORIAL EXPERIMENTS

Three different kinds of analysis are presented here. The first uses data with no replication, but centerpoints. Standard ANOVA can be used here because there is an estimate of within-group variance.

    The second assumes that the entire factorial was replicated. ANOVA can still be used.

    The third assumes that neither replication nor centerpoints are available to supply an estimate of within-group variance. In this case, effects estimates are computed; these can be examined in a probability plot to make a subjective determination of significance.

```
OPTIONS LS=68 PS=48;
/*
        S4_1.SAS—analysis of factorial experiments
    1.  read data and produce three data sets
        D1—one replicate with centerpoints
        D2—two replicates, but no centerpoints
        D3—one replicate without centerpoints
    2.  Use GLM to analyze D1 and D2
    3.  Use GLM to analyze D3 by ignoring some interactions
```

4. Produce effects estimates from D3 which would be used for probability plot analysis

```
*/
DATA D1 D2 D3;
        INFILE 'S4_1.DAT';
        INPUT A B C D Y CP REP;
        IF REP=1 THEN OUTPUT D1;
        IF CP=0 THEN OUTPUT D2;
        IF CP=0 AND REP=1 THEN OUTPUT D3;

*       Use GLM on datasets with error estimates;
PROC GLM DATA=D1;
        CLASS A B C D;
        MODEL Y=A|B|C|D;
        TITLE 'ANOVA for unreplicated factorial with centerpoints';
PROC GLM DATA=D2;
        CLASS A B C D;
        MODEL Y=A|B|C|D;
        TITLE 'ANOVA for replicated factorial without centerpoints';

*       Use GLM on unreplicated factorial by ignoring interactions;
PROC GLM DATA=D2;
        CLASS A B C D;
        MODEL Y=A B C D
            A*B B*C A*D B*C B*D C*D;
        TITLE 'ANOVA with some interactions ignored';

*       Produce effects estimates from D3;
PROC SUMMARY DATA=D3;
        CLASS A B C D;
        VAR Y;
        OUTPUT OUT=D2 MEAN=Y;
DATA D3;
        ARRAY FACTS (4) D C B A;
        LENGTH NAM1 $ 1 NAM2 $ 1 NAM3 $ 1 NAM4 $ 1;
        LENGTH EFFNAME $ 4;
        ARRAY NAMES (4) NAM1-NAM4;
        ARRAY INDX (4) I1-I4;
        RETAIN NAM1 ('D'); RETAIN NAM2 ('C');
        RETAIN NAM3 ('B'); RETAIN NAM4 ('A');
        SET D2;
        EFFECT=_TYPE_;
```

```
        COMPLEX=0;
        EFFNAME= '      ';
        DO I=1 TO 4;
             INDX{I}=MOD(_TYPE_,2);
             COMPLEX=COMPLEX+INDX{I};
             IF INDX{I} EQ 1 THEN SUBSTR (EFFNAME,I,1)=NAMES{I};
             _TYPE_=INT (_TYPE_/2);
        END;
        VAL=1;
        DO I=1 TO 4;
             IF INDX{I}=1 THEN VAL=VAL*FACTS{I}; END;
        IF COMPLEX GT 0 THEN OUTPUT;
        DROP _TYPE_;
PROC SUMMARY DATA=D3 NWAY;
        CLASS EFFECT VAL;
        ID EFFNAME;
        VAR Y;
        OUTPUT OUT=D4 MEAN=Y;
PROC SORT DATA=D4; BY EFFECT VAL;
DATA D5;
        RETAIN OVAL;
        SET D4; BY EFFECT;
        IF FIRST.EFFECT THEN OVAL=Y;
        IF LAST.EFFECT THEN DO;
             MAG=Y-OVAL;
             OUTPUT;END;
        KEEP EFFNAME MAG;
PROC SORT DATA=D5; BY MAG;
PROC PRINT DATA=D5;
        TITLE 'Effects Estimates for Unreplicated Factorial';
```

PROGRAM S4_2.SAS: ANALYSIS OF COVARIANCE

In this program, rinse time is a treatment factor and rinse volume is an uncontrolled covariate. General Linear Models (GLM) is used to determine the significance of rinse time and to estimate regression coefficients for rinse volume. A second analysis examines the regression in detail for each level of rinse time.

```
OPTIONS LS=68 PS=48;
/*
        S4_2.SAS—Analysis of Covariance
   1.  Read raw data. Rinse time is a categorical treatment factor, rinse volume
       is an uncontrolled covariate
```

2. Do the analysis of covariance
3. Do separate regressions by rinse time to see if the rinse volume slopes are the same for both levels of rinse time.

```
*/
DATA D1;
        INFILE 'S4_2.DAT';
        INPUT RTIME RVOL CD;

*       Analysis of Covariance;
PROC GLM DATA=D1;
        CLASS RTIME;
        MODEL CD=RTIME RVOL/SOLUTION;
        TITLE 'Analysis of Covariance';

*       Separate regressions;
PROC GLM DATA=D1;
        MODEL CD=RVOL/SOLUTION;
        BY RTIME;
        TITLE 'Separate regressions by rinse time';
```

PROGRAM S5_1.SAS: FRACTIONAL FACTORIAL ANALYSIS

This program analyzes data from a 2^{5-1} resolution V fractional factorial experiment. In the first analysis, effects estimates are computed. In the second, a factor is excluded from the analysis to provide artificial replication.

```
OPTIONS LS=68 PS=48;
/*
        S5_1.SAS—Analysis of Fractional Factorials
    1.  Read raw data for 2**(5–1) fractional factorial with SPEED as the response
    2.  Compute effects estimates including all two-way interactions
    3.  ANOVA excluding one factor

*/
DATA D1;
        INFILE 'S5_1.DAT';
        INPUT HEAD BACK FOOT HYDR BREA SPEED;

*       Compute effects estimates;
PROC SUMMARY DATA=D1;
        CLASS HEAD BACK FOOT HYDR BREA;
```

```
      VAR SPEED;
      OUTPUT OUT=D2 MEAN=SPEED;

*     COMPLEX indicates level of interaction;
DATA D3;
      ARRAY FACTS(5) BREA HYDR FOOT BACK HEAD;
      LENGTH NAM1 $ 1 NAM2 $ 1 NAM3 $ 1 NAM4 $ 1 NAM5 $ 1;
      LENGTH EFFNAME $ 5;
      ARRAY NAMES(5) NAM1-NAM5;
      ARRAY INDX(5) I1-I5;
      RETAIN NAM1 ('R'); RETAIN NAM2 ('Y');
      RETAIN NAM3 ('F');
      RETAIN NAM4 ('B'); RETAIN NAM5 ('H');
      SET D2;
      EFFECT=_TYPE_;
      COMPLEX=0;
      EFFNAME= '     ';
      DO I=1 TO 5;
            INDX{I}=MOD(_TYPE_,2);
            COMPLEX=COMPLEX+INDX{I};
            IF INDX{I}'
            EQ 1 THEN SUBSTR (EFFNAME,I,1)=NAMES{I};
            _TYPE_=INT(_TYPE_/2);
      END;
      VAL=1;
      DO I=1 TO 5;
            IF INDX{I}=1 THEN VAL=VAL*FACTS{I};
      END;
      IF (COMPLEX GT 0 AND COMPLEX LE 2) THEN OUTPUT;
      DROP _TYPE_;
PROC SUMMARY DATA=D3 NWAY;
      CLASS EFFECT VAL;
      ID EFFNAME;
      VAR SPEED;
      OUTPUT OUT=D4 MEAN=SPEED;
PROC SORT DATA=D4; BY EFFECT VAL;
DATA D5;
      RETAIN OVAL;
      SET D4; BY EFFECT;
      IF FIRST.EFFECT THEN OVAL=SPEED;
      IF LAST.EFFECT THEN DO;
            MAG=SPEED-OVAL;
            OUTPUT;END;
```

```
        KEEP EFFNAME MAG;
PROC SORT DATA=D5; BY MAG;
PROC PRINT DATA=D5;
        TITLE 'Effects estimates for 2** (5-1)';


*       Exclude HYDR and use ANOVA;
PROC GLM DATA=D1;
        CLASS HEAD BACK FOOT HYDR BREA;
        MODEL SPEED=HEAD BACK FOOT BREA
        HEAD*BACK HEAD*FOOT HEAD*BREA
        BACK*FOOT BACK*BREA
        FOOT*BREA;
        TITLE 'ANOVA with HYDR excluded';
```

PROGRAM S5_2.SAS: ANALYSIS OF PLACKETT-BURMAN EXPERIMENT

This program uses GLM to analyze a seven-factor Plackett-Burman experiment with centerpoints. No interactions are tested.

```
OPTIONS LS=68 PS=48;
/*

        S5_2.SAS—Analysis of Plackett Burman
    1.  Read raw data A through G are factors in a seven-factor PB experiment
        with eight runs. T is the response. Six centerpoints are included.
    2.  Use ANOVA to find significant effects. Note that interactions are not
        tested.
*/
DATA D1;
        INFILE 'S5_2.DAT';
        INPUT A B C D E F G T;
PROC GLM DATA=D1;
        CLASS A B C D E F G;
        MODEL T=A B C D E F G;
        TITLE 'Analysis of Plackett Burman Experiment';
```

PROGRAM S6_1.SAS: CONTOUR PLOTS

This program synthesizes data and makes contour plots for several common surfaces.

```
OPTIONS LS=68 PS=24;
/*

        S6_1.SAS—Make contour plots of typical surfaces
    1.  Synthesize data for six typical surfaces
    2.  Make a contour plot for each

*/
DATA D1;
        DO X=-1 TO 2 BY .105; DO Y=-2 TO 2 BY .203;
                Z6=18-(X+Y+.34)**2;
                Z7=38-.25*(X+1.4)**2+.18*(1.5*X+Y-1)**2+12;
                Z8=2*(9-X**2-Y**2);
                Z9=(SQRT(3-X**2/4.1+Y**2/8) -0.3)*27;
                Z10=MAX(X**2*Y/(X**4+Y++2)+.3,0)*27;
                Z11=(COS(3*(X+2))*COS(2.4*(Y+2))*
                EXP(-1.0*SQRT(X**2+Y**2))+ 0.1)*18;
        OUTPUT;END;END;
*       Set up graphics;
SYMBOL1 L=1; SYMBOL2 L=3; SYMBOL3 L=5; SYMBOL4 L=6;
SYMBOL5 L=8; SYMBOL6 L=10; SYMBOL7 L=12;
FILENAME GFIL 'S6_1A.GSF';
GOPTIONS DEVICE='HP7475'
        VPOS=48 HPOS=60 ASPECT=1 HBY=1.5
        HTITLE=5 FTITLE=SIMPLEX
        HTEXT=2 FTEXT=SIMPLEX
        ROTATE=LANDSCAPE
        NODISPLAY
        GSFMODE=REPLACE
        GSFNAME=GFIL;
*       Make the plots;
PROC GCONTOUR DATA=D1;
        PLOT Y*X=Z6/JOINLLEVELS=1,3,5,6,8,10,12;
        TITLE 'Stationary Ridge';
        RUN;
FILENAME GFIL 'S6_1B.GSF';
PROC GCONTOUR DATA=D1;
        PLOT Y*X=Z7/JOINLLEVELS=1,3,5,6,8,10,12;
        TITLE 'Rising Ridge';
        RUN;
FILENAME GFIL '6_1C.GSF';
PROC GCONTOUR DATA=D1;
        PLOT Y*X=Z8/JOINLLEVELS=1,3,5,6,8,10,12;
        TITLE 'Simple Maximum';
        RUN;
```

```
FILENAME GFIL '6_1D.GSF';
PROC GCONTOUR DATA=D1;
        PLOT Y*X=Z9/JOINLLEVELS=1,3,5,6,8,10,12;
        TITLE 'Saddle Point';
        RUN;
FILENAME GFIL '6_1E.GSF';
PROC GCONTOUR DATA=D1;
        PLOT Y*X=Z10/JOINLLEVELS=1,3,5,6,8,10,12;
        TITLE 'Cliff';
        RUN;
FILENAME GFIL '6_1F.GSF';
PROC GCONTOUR DATA=D1;
        PLOT Y*X=Z11/JOINLLEVELS=1,3,5,6,8,10,12;
        TITLE 'Multiple Optima';
        RUN;
```

PROGRAM 6_2.SAS: CENTRAL COMPOSITE ANALYSIS

This program uses PROC RSREG to analyze data from a central composite design based on the 2^{5-1} fractional factorial. The LACKFIT option tests for lack of fit.

```
OPTIONS LS=68 PS=48;
/*
        6_2.SAS—Analysis of central composite designs

    1.  Read raw data and scale some parameters RHO is the response in a five-
        factor central composite
    2.  Use PROC RSREG to estimate parameters in a quadratic model. Include a
        lack of fit test.

*       Read the raw data;
DATA D1;
        INFILE 'S6_2.DAT';
        INPUT TEMP FLOW PRESS ClF POWER RHO;
        PRESS=PRESS/5; ClF=ClF/6.25; POWER=POWER/3.75;
*       Estimate the response surface;
PROC RSREG DATA=D1 OUT=D1A;
        MODEL RHO=TEMP FLOW PRESS ClF POWER
                /LACKFIT NOOPT RESIDUAL;
```

Table A.1: Upper 5% Percentiles of the F-distribution. Numerator degrees of freedom (v_1) are shown in columns; denominator degrees of freedom (v_2) are found in rows. Hence, $F_{4,6,0.95} = 4.53$.

v_1

v_2	1	2	3	4	5	6	7	8	9	10	12	15	20	24	30	40	60	120
1	161.5	199.5	215.7	224.6	230.2	234.0	236.8	238.9	240.5	241.9	243.9	246.0	248.0	249.1	250.1	251.1	252.2	253.3
2	18.51	19.00	19.16	19.25	19.30	19.33	19.35	19.37	19.38	19.40	19.41	19.43	19.45	19.45	19.46	19.47	19.48	19.49
3	10.13	9.55	9.28	9.12	9.01	8.94	8.89	8.85	8.81	8.79	8.74	8.70	8.66	8.64	8.62	8.59	8.57	8.55
4	7.71	6.94	6.59	6.39	6.26	6.16	6.09	6.04	6.00	5.96	5.91	5.86	5.80	5.77	5.75	5.72	5.69	5.66
5	6.61	5.79	5.41	5.19	5.05	4.95	4.88	4.82	4.77	4.74	4.68	4.62	4.56	4.53	4.50	4.46	4.43	4.40
6	5.99	5.14	4.76	4.53	4.39	4.28	4.21	4.15	4.10	4.06	4.00	3.94	3.87	3.84	3.81	3.77	3.74	3.70
7	5.59	4.74	4.35	4.12	3.97	3.87	3.79	3.73	3.68	3.64	3.57	3.51	3.44	3.41	3.38	3.34	3.30	3.27
8	5.32	4.46	4.07	3.84	3.69	3.58	3.50	3.44	3.39	3.35	3.28	3.22	3.15	3.12	3.08	3.04	3.01	2.97
9	5.12	4.26	3.86	3.63	3.48	3.37	3.29	3.23	3.18	3.14	3.07	3.01	2.94	2.90	2.86	2.83	2.79	2.75
10	4.96	4.10	3.71	3.48	3.33	3.22	3.14	3.07	3.02	2.98	2.91	2.85	2.77	2.74	2.70	2.66	2.62	2.58
11	4.84	3.98	3.59	3.36	3.20	3.09	3.01	2.95	2.90	2.85	2.79	2.72	2.65	2.61	2.57	2.53	2.49	2.45
12	4.75	3.89	3.49	3.26	3.11	3.00	2.91	2.85	2.80	2.75	2.69	2.62	2.54	2.51	2.47	2.43	2.38	2.34
13	4.67	3.81	3.41	3.18	3.03	2.92	2.83	2.77	2.71	2.67	2.60	2.53	2.46	2.42	2.38	2.34	2.30	2.25
14	4.60	3.74	3.34	3.11	2.96	2.85	2.76	2.70	2.65	2.60	2.53	2.46	2.39	2.35	2.31	2.27	2.22	2.18
15	4.54	3.68	3.29	3.06	2.90	2.79	2.71	2.64	2.59	2.54	2.48	2.40	2.33	2.29	2.25	2.20	2.16	2.11
16	4.49	3.63	3.24	3.01	2.85	2.74	2.66	2.59	2.54	2.49	2.42	2.35	2.28	2.24	2.19	2.15	2.11	2.06
17	4.45	3.59	3.20	2.96	2.81	2.70	2.61	2.55	2.49	2.45	2.38	2.31	2.23	2.19	2.15	2.10	2.06	2.01
18	4.41	3.55	3.16	2.93	2.77	2.66	2.58	2.51	2.46	2.41	2.34	2.27	2.19	2.15	2.11	2.06	2.02	1.97
19	4.38	3.52	3.13	2.90	2.74	2.63	2.54	2.48	2.42	2.38	2.31	2.23	2.16	2.11	2.07	2.03	1.98	1.93
20	4.35	3.49	3.10	2.87	2.71	2.60	2.51	2.45	2.39	2.35	2.28	2.20	2.12	2.08	2.04	1.99	1.95	1.90
21	4.32	3.47	3.07	2.84	2.68	2.57	2.49	2.42	2.37	2.32	2.25	2.18	2.10	2.05	2.01	1.96	1.92	1.87
22	4.30	3.44	3.05	2.82	2.66	2.55	2.46	2.40	2.34	2.30	2.23	2.15	2.07	2.03	1.98	1.94	1.89	1.84
23	4.28	3.42	3.03	2.80	2.64	2.53	2.44	2.37	2.32	2.27	2.20	2.13	2.05	2.01	1.96	1.91	1.86	1.81
24	4.26	3.40	3.01	2.78	2.62	2.51	2.42	2.36	2.30	2.25	2.18	2.11	2.03	1.98	1.94	1.89	1.84	1.79
25	4.24	3.39	2.99	2.76	2.60	2.49	2.40	2.34	2.28	2.24	2.16	2.09	2.01	1.96	1.92	1.87	1.82	1.77
26	4.23	3.37	2.98	2.74	2.59	2.47	2.39	2.32	2.27	2.22	2.15	2.07	1.99	1.95	1.90	1.85	1.80	1.75
27	4.21	3.35	2.96	2.73	2.57	2.46	2.37	2.31	2.25	2.20	2.13	2.06	1.97	1.93	1.88	1.84	1.79	1.73
28	4.20	3.34	2.95	2.71	2.56	2.45	2.36	2.29	2.24	2.19	2.13	2.04	1.96	1.91	1.87	1.82	1.77	1.71
29	4.18	3.33	2.93	2.70	2.55	2.43	2.35	2.28	2.22	2.18	2.10	2.03	1.94	1.90	1.85	1.81	1.75	1.70
30	4.17	3.32	2.92	2.69	2.53	2.42	2.33	2.27	2.21	2.16	2.09	2.01	1.93	1.89	1.84	1.79	1.74	1.68
40	4.08	3.23	2.84	2.61	2.45	2.34	2.25	2.18	2.12	2.08	2.00	1.92	1.84	1.79	1.74	1.69	1.64	1.58
60	4.00	3.15	2.76	2.53	2.37	2.25	2.17	2.10	2.04	1.99	1.92	1.84	1.75	1.70	1.65	1.59	1.53	1.47
120	3.92	3.07	2.68	2.45	2.29	2.18	2.09	2.02	1.96	1.91	1.3	1.75	1.66	1.61	1.55	1.50	1.43	1.35

INDEX